CRC Handbook High Resolution Infrared Laboratory Spectra of Atmospheric Interest

Editors

David G. Murcray
Department of Physics
University of Denver
Denver, Colorado

Aaron Goldman
Department of Physics
University of Denver
Denver, Colorado

CRC Press
Taylor & Francis Group
Boca Raton London New York

CRC Press is an imprint of the
Taylor & Francis Group, an **informa** business

PREFACE

Part of the atmospheric spectroscopy research at the University of Denver consists of detailed identification of the atmospheric absorption features present in high resolution infrared solar spectra obtained from various altitudes. This study revealed a lack of laboratory data for many of the compounds of interest with sufficiently high resolution to be used for identification. Therefore, a laboratory study was undertaken to obtain high resolution spectra of many of the species of interest. The spectra presented in this Handbook are the result of this study.

A number of individuals participated in the preparation of this handbook, most notably F. S. Bonomo, C. M. Bradford, G. R. Cook, F. H. Murcray, F. J. Murcray, and J. W. VanAllen. The research was supported by the Chemical Manufacturers Association and the National Science Foundation.

THE EDITORS

Professor David G. Murcray has been engaged in atmospheric research with particular emphasis on atmospheric spectroscopy for over 20 years. He received his B.S. from the University of Denver, his M.S. from Oklahoma State University, and his Ph.D. from the University of Denver.

Professor Murcray began his career as a teaching fellow at Harvard University from 1948 to 1949 and at Oklahoma State University from 1949 to 1950. He was an assistant instructor at Kansas University from 1950 to 1951 and a research mathematician for Phillips Petroleum Company the following year. Since 1952 he has been at the University of Denver, where he is currently Professor of Physics.

His atmospheric research has included measurements of water vapor and other minor constituents and trace constituents in the region from 13 km to 40 km, using spectroscopic techniques. This research led to the initial detection of HNO_3 and NO_2 in the lower stratosphere. Included in his research interests are atmospheric spectroscopy (particularly in the infrared), infrared instrumentation, Fourier transform spectroscopy and atmospheric chemistry.

A member of Phi Beta Kappa, Sigma Pi Sigma, Sigma Xi, and a Fellow of the Optical Society, Professor Murcray has published numerous papers and journal articles concerned with atmospheric spectroscopy and atmospheric composition.

Aaron Goldman, Research Professor of Physics at the University of Denver, has done extensive research in quantitative molecular and atomic spectroscopy since receiving his M.Sc. in 1962 from the Hebrew University, Jerusalem, Israel. He received his D.Sc. in 1965 from the Technion-Israel Institute of Technology, Haifa, Israel, where he was involved in high temperature spectral emissivity infrared and ultraviolet laboratory studies of atmospheric gases. From 1966 to present, his research has been concentrated on quantitative analysis of infrared and ultraviolet absorption of solar radiation by the earth's atmospheric gases, atomic and molecular solar lines, and infrared emission by the earth's atmosphere, from high resolution spectra obtained with ground-based and air-borne spectrometers. As part of this research, Dr. Goldman demonstrated the detection and quantification of several new molecular species in the lower stratosphere. A member of Sigma Xi and the American Geophysical Union, Dr. Goldman has authored or co-authored over 100 papers and journal articles related to atmospheric and solar spectroscopy.

TABLE OF CONTENTS

INTRODUCTION

Infrared spectroscopy has been used for many years as a means of identifying and quantifying the constituents present in gaseous samples. As a result, a large number of infrared spectra have been published either in the form of handbooks or index cards (many of them by the chemical industry). For most industrial applications, resolution is not crucial, and the majority of spectra presented in previous publications have moderate to low resolution (i.e., not better than 1 cm^{-1}).

During the last decade, infrared spectroscopy also became a powerful tool in atmospheric studies. Locating a source at some distance from the spectrometer with the intervening atmosphere as an absorber provides spectra which contain a wealth of information on atmospheric composition. This technique is used extensively to obtain information concerning gases present in natural and polluted atmospheres. Recent studies in atmospheric chemistry, particularly those associated with the ozone layer, have shown that very small amounts of certain compounds can act as catalysts and have a significant impact on atmospheric photochemistry and composition. As a result, there is considerable current interest in obtaining data concerning the concentration of a large number of compounds which are present in the atmosphere at very low levels. This leads to the need for an increase in sensitivity of measurement techniques, including infrared spectroscopy. Increased spectral resolution improves the sensitivity of infrared spectroscopic measurements for almost all species.

One of the major difficulties encountered in attempting to use infrared spectroscopy for atmospheric measurements is the interference from absorption features due to other atmospheric constituents. The common minor constituents (H_2O, CO_2, N_2O, CH_4, O_3, and CO) all have strong infrared absorption features and these tend to produce absorption lines that interfere with those due to the molecular species of interest. Detection and measurement of gases in atmospheric spectra must be accomplished against this interference, and increased spectral resolution becomes very important in achieving detection under such conditions.

A recent publication by Graedel[1] lists 1600 gases which are known or thought to be present in the atmosphere. Assembling high resolution spectra of all these gases is obviously beyond the scope of this Handbook. Spectra in this Handbook represent a relatively small number of these gases, the majority of which were chosen because of current interest in the stratospheric ozone layer. Most of the spectra presented are in the 5 to 15 μ region, where most of our recent atmospheric spectral studies were carried out. Many of the molecules of interest also have strong bands at shorter and longer wavelengths which are not yet included here, but which will hopefully be included in future editions of this Handbook.

The present spectra are at least one order of magnitude higher resolution than other comparably extensive compilations of spectra. The development of modern laser spectrometers allows still higher resolution over limited spectral ranges. The resolution achieved in the spectra presented here is close to the limit for which it is practical to present spectra in handbook form. While workers often exchange data on magnetic tape, it has been our experience that a great deal of preliminary work can be accomplished by using a hard copy of the appropriate spectra. It is on the basis of our own use of hard copy spectra that we felt it worthwhile to publish this Handbook.

DESCRIPTION OF THE ATLAS

The spectra presented in this Handbook were obtained for either of two objectives. The first major objective was possible detection and measurement of new constituents in the earth's atmosphere. In most cases, the expected atmospheric absorptions from such a constituent will be quite weak. Therefore, the amount of gas in the absorption cell was adjusted so that the peak absorption in the spectrum was less than 50%. However, in some cases portions of the band may be strong enough that this criterion was exceeded.

The second objective was to obtain data for some of the common minor species (such as CH_4, N_2O, etc.) with large gas amounts in the absorption cell. Such data are needed for identifying absorption features which occur in atmospheric spectra taken over long atmospheric paths, but which are not adequately documented in previous studies. In some cases we have included spectra taken with more than one gas in the optical path for spectral regions with overlapping absorption features.

All of the spectra were taken using Fourier Transform Spectrometer systems. These systems are gradually replacing grating spectrometers for wide spectral regions in most laboratories since they have a number of advantages over grating spectrometers. One of the major advantages is the precise frequency data (linearity as well as absolute frequency calibration) that can be obtained. In many cases presented here, the spectra were calibrated against suitable reference absorption lines to within 0.1 of a resolution element. In all cases the frequencies are accurate to within a resolution element. Some of the spectra were ratioed by a corresponding spectrum of an empty cell to remove envelope trends. Some of these ratioed spectra can be considered as approximate transmittance.

The spectrum for each molecule is plotted on two scales. The first is a condensed scale, typically covering a span of 75 to 300 cm^{-1} in one frame. The second scale is an expanded view covering 20 or 10 cm^{-1} per frame. This was necessary to permit presentation of an entire band, while at the same time preserving details inherent in the high resolution used. If no high resolution structure is apparent, the expanded plots are not included.

Each plot contains a title with the name of the molecule, its chemical formula, an alternate name (if appropriate), the gas pressure, cell length and estimated resolution of the spectrum. With the two different instruments used, resolutions of 0.02 cm^{-1} and 0.06 cm^{-1} are present. All of the spectra were obtained under room conditions. In addition, each condensed scale frame contains a few additional explanatory notes. These notes typically indicate the presence of residual H_2O or CO_2, filter structure, or channel spectra.

The plots are ordered in the List of Molecular Species by formula according to an alphabetical scheme (Herzberg,[2] Rao and Mathews[3]).

A complete cross-reference index, alphabetized by molecule name, is also included. This contains references to alternate names and formula presentations that may be more familiar to the reader.

ACKNOWLEDGMENTS

Acknowledgment is made to the National Center for Atmospheric Research, which is sponsored by the National Science Foundation, for computer time used in this research. The computer programs used for plotting the spectra were written by Darwin Rolens. The figures were carefully plotted and prepared by Barbra Cox.

REFERENCES

1. **Graedel, T. E.,** *Chemical Compounds in the Atmosphere,* Academic Press, New York, 1978.
2. **Herzberg, G.,** *Molecular Spectra and Molecular Structure* II. *Infrared and Raman Spectra of Polyatomic Molecules,* D. Van Nostrand Company, Princeton, New Jersey, 1962, 568.
3. **Rao, K. N. and Mathews, C. W.,** *Molecular Spectroscopy: Modern Research,* Academic Press, New York, 1972, 412.

LIST OF MOLECULAR SPECIES

CROSS INDEX

	Molecule	Plot

Name	Acetylene	
Alternate name	Ethyne	
Index reference	C_2H_2	
Alternate formula	C_2H_2	
Plots	680-840cm^{-1} 8.0 Torr 10cm R $= 0.06$cm^{-1}	(130)

Name	Ammonia	
Alternate name	Ammonia	
Index reference	NH_3	
Alternate formula	NH_3	
Plots	750-950cm^{-1} 0.5 Torr 10cm R $= 0.06$cm^{-1}	(290A)
	1070-210cm^{-1} 5.0 Torr 5cm R $= 0.06$cm^{-1}	(290B)

Name	Carbon disulfide	
Alternate name	Carbon disulfide	
Index reference	CS_2	
Alternate formula	CS_2	
Plots	1460-1600cm^{-1} 0.5 Torr 10cm R $= 0.06$cm^{-1}	(260)

Name	Carbon oxysulfide	
Alternate name	See Carbonyl sulfide	
Index reference		
Alternate formula		
Plots		

Name	Carbonyldichloride, Carbonyl chloride	
Alternate name	Phosgene, chloroformylchloride	
Index reference	CCl_2O	
Alternate formula	$COCl_2$	
Plots	800-880cm^{-1} 0.8 Torr 10cm R $= 0.06$cm^{-1}	(70)

Name	Carbonyl chlorofluoride	
Alternate name	Chloroformylfluoride	
Index reference	$CClFO$	
Alternate formula	$COClF$	
Plots	1070-1150cm^{-1} 0.34 Torr 10cm R $= 0.06$cm^{-1}	(60)

Name	Carbonyl difluoride, Carbonyl fluoride	
Alternate name	Fluoroformylfluoride	
Index reference	CF_2O	
Alternate formula	COF_2	
Plots	740-820cm^{-1} 3.1 Torr 10cm R $= 0.06$cm^{-1}	(100A)
	1180-1280cm^{-1} 3.1 Torr 10cm R $= 0.06$cm^{-1}	(100B)

CROSS INDEX (continued)

	Molecule	Plot

Name	Carbonyl sulfide	
Alternate name	Carbon oxysulfide	
Index reference	COS	
Alternate formula	OCS	
Plots	2000-2120cm^{-1} 2.0 Torr 10cm R = 0.06cm^{-1}	(250)

Name	Chlorinenitrate	
Alternate name	Chlorine nitrate	
Index reference	ClONO$_2$	
Alternate formula	ClONO$_2$	
Plots	770-850cm^{-1} 14.0 Torr 5cm R = 0.06cm^{-1}	(230B)
	770-850cm^{-1} 1.9 Torr 10cm R = 0.06cm^{-1}	(230A)
	1255-1355cm^{-1} 14.0 Torr 5cm R = 0.06cm^{-1}	(230C)

Name	Chlorine nitrate + nitric acid	
Alternate name	Chlorine nitrate + nitric acid	
Index reference	ClONO$_3$, HNO$_3$	
Alternate formula	ClONO$_3$, HNO$_3$	
Plots	1260-1340cm^{-1} 1.2 Torr 5cm R = 0.06cm^{-1}	(240)

Name	Chlorodifluoromethane	
Alternate name	F-22	
Index reference	CHClF$_2$	
Alternate formula	CHF$_2$Cl	
Plots	770-850cm^{-1} 3.0 Torr 5cm R = 0.06cm^{-1}	(200)

Name	Chloroformylchloride
Alternate name	See Carbonyl dichloride
Index reference	
Alternate formula	
Plots	

Name	Chloroformylfluoride
Alternate name	See Chlorofluoride
Index reference	
Alternate formula	
Plots	

Name	Chloromethane
Alternate name	See Methyl chloride
Index reference	
Alternate formula	
Plots	

CROSS INDEX (continued)

	Molecule	Plot

Name	Chlorotrifluoromethane	
Alternate name	F-13	
Index reference	$CClF_3$	
Alternate formula	CF_3Cl	
Plots	1070-1250cm^{-1} 0.4 Torr 10cm R = 0.06cm^{-1}	(10)

Name	Dichlorodifluoromethane	
Alternate name	F-12	
Index reference	CCl_2F_2	
Alternate formula	CF_2Cl_2	
Plots	840-940cm^{-1} 4.1 Torr 5cm R = 0.06cm^{-1}	(20A)
	1070-1250cm^{-1} 0.4 Torr 10cm R = 0.06cm^{-1}	(20B)
	1080-1180cm^{-1} 4.1 Torr 5cm R = 0.06cm^{-1}	(20C)

Name	1,2-Dichloroethane	
Alternate name	Ethylene chloride	
Index reference	$C_2H_4Cl_2$	
Alternate formula	$C_2H_4Cl_2$	
Plots	700-760cm^{-1} 3.5 Torr 10cm R = 0.06cm^{-1}	(190A)
	1200-1300cm^{-1} 3.5 Torr 10cm R = 0.06cm^{-1}	(190B)

Name	Dichlorofluoromethane	
Alternate name	F-21	
Index reference	$CHCl_2F$	
Alternate formula	$CHFCl_2$	
Plots	700-860cm^{-1} 5.0 Torr 5cm R = 0.06cm^{-1}	(210A)
	1060-1300cm^{-1} 5.0 Torr 5cm R = 0.06cm^{-1}	(210B)

Name	Dichloromethane	
Alternate name	Methylene chloride	
Index reference	CH_2Cl_2	
Alternate formula	CH_2Cl_2	
Plots	720-800cm^{-1} 5.9 Torr 5cm R = 0.06cm^{-1}	(150A)
	1240-1300cm^{-1} 5.9 Torr 5cm R = 0.06cm^{-1}	(150B)

Name	1,2-Dichlorotetrafluoroethane	
Alternate name	F-114	
Index reference	$C_2Cl_2F_4$	
Alternate formula	$C_2F_4Cl_2$	
Plots	810-950cm^{-1} 0.9 Torr 10cm R = 0.06cm^{-1}	(40A)
	1080-1300cm^{-1} 0.9 Torr 10cm R = 0.06cm^{-1}	(40B)

Name	Dinitrogen pentoxide	
Alternate name	Nitrogen pentoxide	
Index reference	N_2O_5	
Alternate formula	N_2O_5	
Plots	1220-1270cm^{-1} 5.0 Torr 5cm R = 0.06cm^{-1}	(320)

CROSS INDEX (continued)

	Molecule	Plot
Name	Ethane	
Alternate name	Ethane	
Index reference	C_2H_6	
Alternate formula	C_2H_6	
Plots	780-880cm^{-1} 3.9 Torr 10cm R = 0.06cm^{-1}	(140A)
	2840-3100cm^{-1} 3.9 Torr 10cm R = 0.06cm^{-1}	(140B)
Name	Ethylene chloride	
Alternate name	See 1,2-Dichloroethane	
Index reference		
Alternate formula		
Plots		
Name	Ethylene trichloride	
Alternate name	See Trichloroethylene	
Index reference		
Alternate formula		
Plots		
Name	Ethyne	
Alternate name	See Acetylene	
Index reference		
Alternate formula		
Plots		
Name	Fluoroformylfluoride	
Alternate name	See Carbonyl difluoride	
Index reference		
Alternate formula		
Plots		
Name	Formic acid	
Alternate name	Methanoic acid	
Index reference	CH_2O_2	
Alternate formula	HCOOH	
Plots	1070-1170cm^{-1} 1.6 Torr 10cm R = 0.06cm^{-1}	(220)
Name	F-11	
Alternate name	See Trichlorofluoromethane	
Index reference		
Alternate formula		
Plots		

CROSS INDEX (continued)

Molecule   Plot

Name	F-113
Alternate name	See 1,1,2-Trichlorotrifluoroethane
Index reference	
Alternate formula	
Plots	

Name	F-114
Alternate name	See 1,2-Dichlorotetrafluoroethane
Index reference	
Alternate formula	
Plots	

Name	F-12
Alternate name	See Dichlorodifluoromethane
Index reference	
Alternate formula	
Plots	

Name	F-13
Alternate name	See Chlorotrifluoromethane
Index reference	
Alternate formula	
Plots	

Name	F-14
Alternate name	See Tetrafluoromethane
Index reference	
Alternate formula	
Plots	

Name	F-21
Alternate name	See Dichlorofluoromethane
Index reference	
Alternate formula	
Plots	

Name	F-22
Alternate name	See Chlorodifluoromethane
Index reference	
Alternate formula	
Plots	

Name	Hydrogen nitrate
Alternate name	See Nitric acid
Index reference	
Alternate formula	
Plots	

CROSS INDEX (continued)

	Molecule	Plot

Name	Hydrogen peroxide	
Alternate name	Hydrogen peroxide	
Index reference	H_2O_2	
Alternate formula	H_2O_2	
Plots	1210-1310cm^{-1} 4.0 Torr 10cm R = 0.06cm^{-1}	(280A)
	1200-1340cm^{-1} 2.5 Torr 45cm R = 0.02cm^{-1}	(280B)

Name	Methane	
Alternate name	Methane	
Index reference	CH_4	
Alternate formula	CH_4	
Plots	1200-1400cm^{-1} 10.0 Torr 45cm R = 0.02cm^{-1}	(110B)
	1200-1400cm^{-1} 80.0 Torr 45cm R = 0.02cm^{-1}	(110C)
	1400-1600cm^{-1} 80.0 Torr 45cm R = 0.02cm^{-1}	(110D)
	1600-1800cm^{-1} 80.0 Torr 45cm R = 0.02cm^{-1}	(110E)

Name	Methane	
Alternate name	Methane	
Index reference	CH_4	
Alternate formula	CH_4	
Plots	1210-1410cm^{-1} 20.0 Torr 10cm R = 0.06cm^{-1}	(110A)

Name	Methane + dinitrogen oxide	
Alternate name	Methane + nitrous oxide	
Index reference	CH_4, N_2O	
Alternate formula	CH_4, N_2O	
Plots	1120-1300cm^{-1} 76.0 Torr 5cm R = 0.06cm^{-1}	(120)

Name	Methanoic acid	
Alternate name	See Formic acid	
Index reference		
Alternate formula		
Plots		

Name	Methyl chloride	
Alternate name	Chloromethane	
Index reference	CH_3Cl	
Alternate formula	CH_3Cl	
Plots	670-770cm^{-1} 72.0 Torr 5cm R = 0.06cm^{-1}	(160A)
	950-1110cm^{-1} 72.0 Torr 5cm R = 0.06cm^{-1}	(160B)

Name	Methyl chloroform	
Alternate name	See 1,1,1-Trichloroethane	
Index reference		
Alternate formula		
Plots		

CROSS INDEX (continued)

	Molecule	Plot

Name | Methylene chloride
Alternate name | See Dichloromethane
Index reference |
Alternate formula |
Plots |

Name | Nitric acid
Alternate name | Hydrogen nitrate
Index reference | HNO_3
Alternate formula | HNO_3
Plots | 1270-1370cm^{-1} 0.42 Torr 45cm R = 0.02cm^{-1} | (270D)
| 1670-1750cm^{-1} 0.42 Torr 45cm R = 0.02cm^{-1} | (270E)

Name | Nitric acid
Alternate name | Hydrogen nitrate
Index reference | HNO_3
Alternate formula | HNO_3
Plots | 760-920cm^{-1} 1.57 Torr 10cm R = 0.06cm^{-1} | (270A)
| 840-940cm^{-1} 2.2 Torr 5cm R = 0.06cm^{-1} | (270B)
| 1275-1355cm^{-1} 2.1 Torr 10cm R = 0.06cm^{-1} | (270C)

Name | Nitric oxide
Alternate name | Mononitrogen monoxide
Index reference | NO
Alternate formula | NO
Plots | 1800-1960cm^{-1} 36.0 Torr 5cm R = 0.06cm^{-1} | (300)

Name | Nitrogen dioxide
Alternate name | Nitrogen dioxide
Index reference | NO_2
Alternate formula | NO_2
Plots | 1500-1700cm^{-1} 2.0 Torr 45cm R = 0.02cm^{-1} | (310)

Name | Nitrogen pentoxide
Alternate name | See Dinitrogen pentoxide
Index reference |
Alternate formula |
Plots |

Name | Phosgene
Alternate name | See Carbonyl chloride
Index reference |
Alternate formula |
Plots |

CROSS INDEX (continued)

	Molecule	Plot

Name	Tetrafluoromethane	
Alternate name	F-14	
Index reference	CF_4	
Alternate formula	CF_4	
Plots	1200-1300cm^{-1} 0.075 Torr 2cm R = 0.06cm^{-1}	(80A)
	1250-1310cm^{-1} 0.04 Torr 5cm R = 0.02cm^{-1}	(80B)

Name	Tetrafluoromethane + methane + nitrous oxide	
Alternate name	F-14 + methane + nitrous oxide	
Index reference	CF_4, CH_4, N_2O	
Alternate formula	CF_4, CH_4, N_2O	
Plots	1050-1350cm^{-1} 0.04, 75 Torr 2.5cm R = 0.02cm^{-1}	(90)

Name	1,1,1-Trichloroethane	
Alternate name	Methyl chloroform	
Index reference	$C_2H_3Cl_3$	
Alternate formula	$C_2H_3Cl_3$	
Plots	700-760cm^{-1} 1.1 Torr 10cm R = 0.06cm^{-1}	(180)

Name	Trichloroethylene	
Alternate name	Ethylene trichloride	
Index reference	C_2HCl_3	
Alternate formula	C_2HCl_3	
Plots	740-900cm^{-1} 2.8 Torr 10cm R = 0.06cm^{-1}	(170)

Name	Trichlorofluoromethane	
Alternate name	F-11	
Index reference	CCl_3F	
Alternate formula	$CFCl_3$	
Plots	810-870cm^{-1} 0.07 Torr 10cm R = 0.06cm^{-1}	(30A)
	810-890cm^{-1} 2.6 Torr 5cm R = 0.06cm^{-1}	(30B)
	1060-1160cm^{-1} 2.6 Torr 5cm R = 0.06cm^{-1}	(30C)

Name	1,1,2-Trichlorotrifluoroethane	
Alternate name	F-113	
Index reference	$C_2Cl_3F_3$	
Alternate formula	$C_2F_3Cl_3$	
Plots	760-940cm^{-1} 2.9 Torr 10cm R = 0.06cm^{-1}	(50A)
	1090-1250cm^{-1} 2.9 Torr 10cm R = 0.06cm^{-1}	(50B)

Spectra

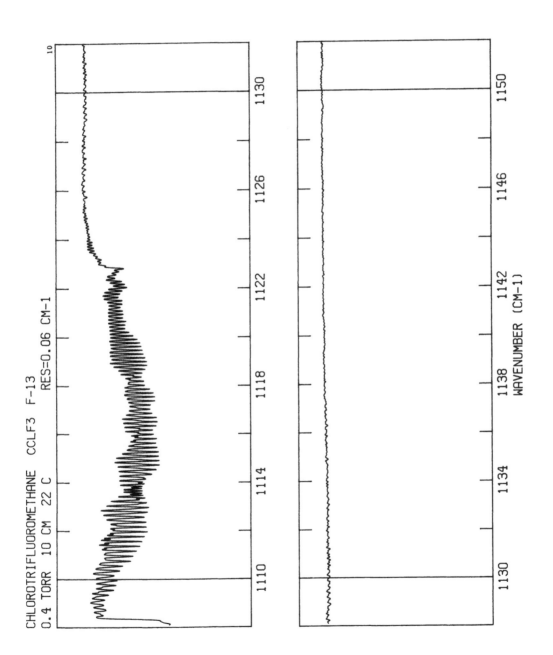

CHLOROTRIFLUOROMETHANE CCLF3 F-13
0.4 TORR 10 CM 22 C   RES=0.06 CM-1

WAVENUMBER (CM-1)

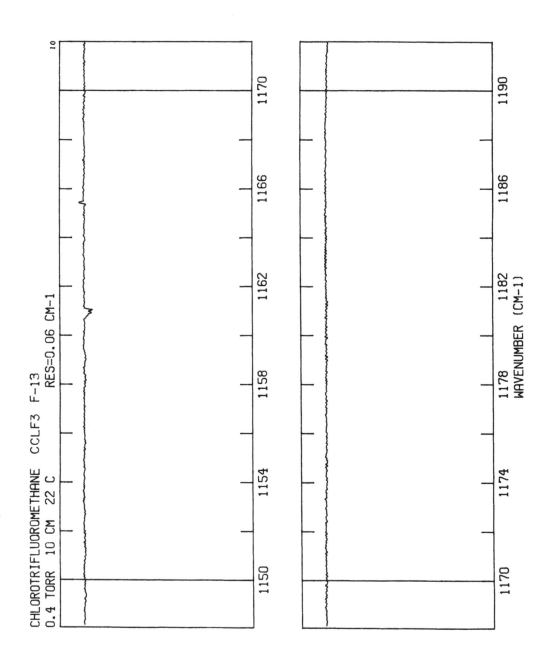

CHLOROTRIFLUOROMETHANE CCLF3 F-13   RES=0.06 CM-1
0.4 TORR 10 CM 22 C

WAVENUMBER (CM-1)

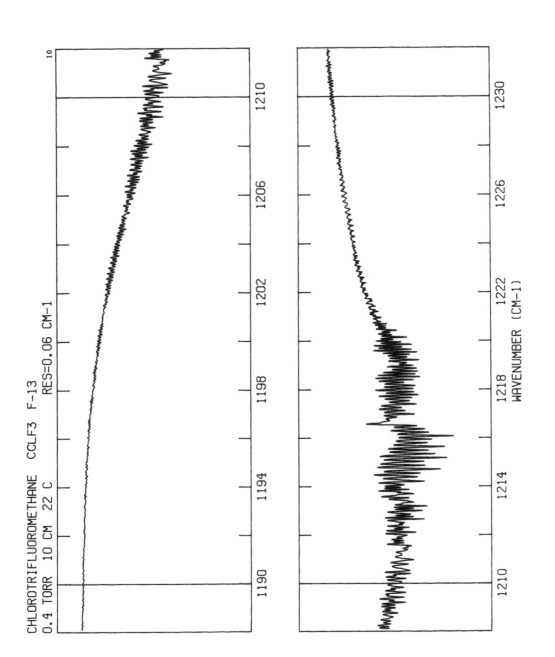

CHLOROTRIFLUOROMETHANE   CCLF3   F-13
0.4 TORR   10 CM   22 C   RES=0.06 CM-1

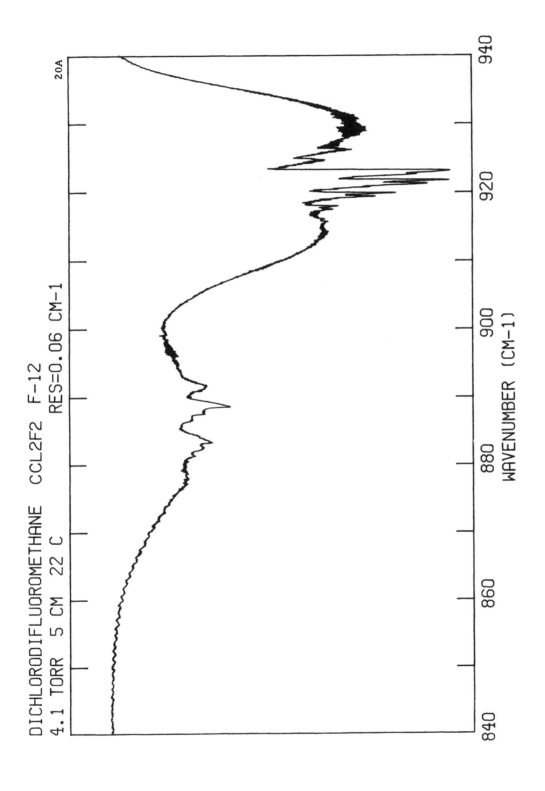

DICHLORODIFLUOROMETHANE CCL2F2 F-12
4.1 TORR 5 CM 22 C RES=0.06 CM-1

20A

WAVENUMBER (CM-1)

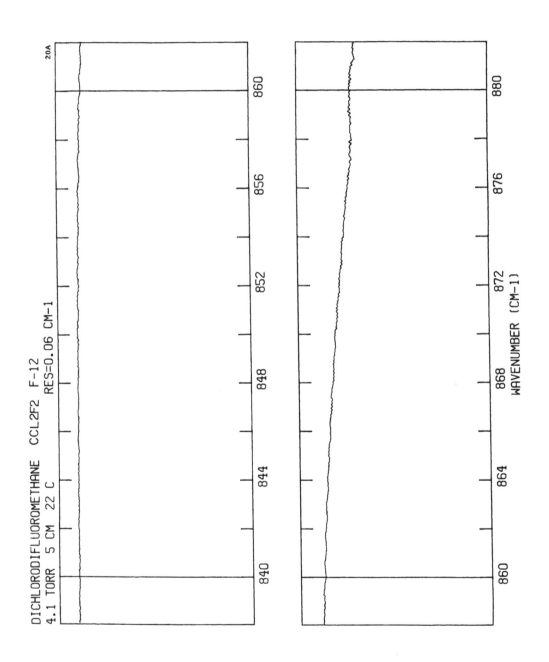

DICHLORODIFLUOROMETHANE CCL2F2 F-12
4.1 TORR 5 CM 22 C RES=0.06 CM-1

WAVENUMBER (CM-1)

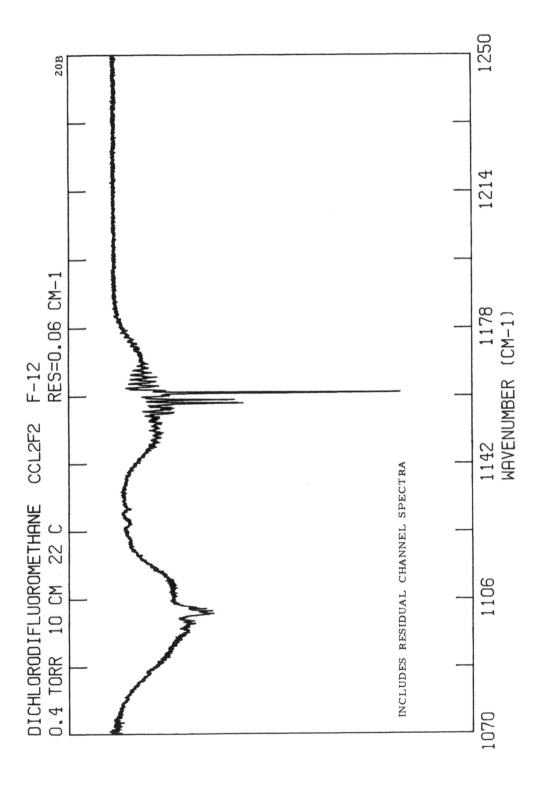

DICHLORODIFLUOROMETHANE CCL2F2 F-12
0.4 TORR 10 CM 22 C   RES=0.06 CM-1

20B

INCLUDES RESIDUAL CHANNEL SPECTRA

WAVENUMBER (CM-1)

1070   1106   1142   1178   1214   1250

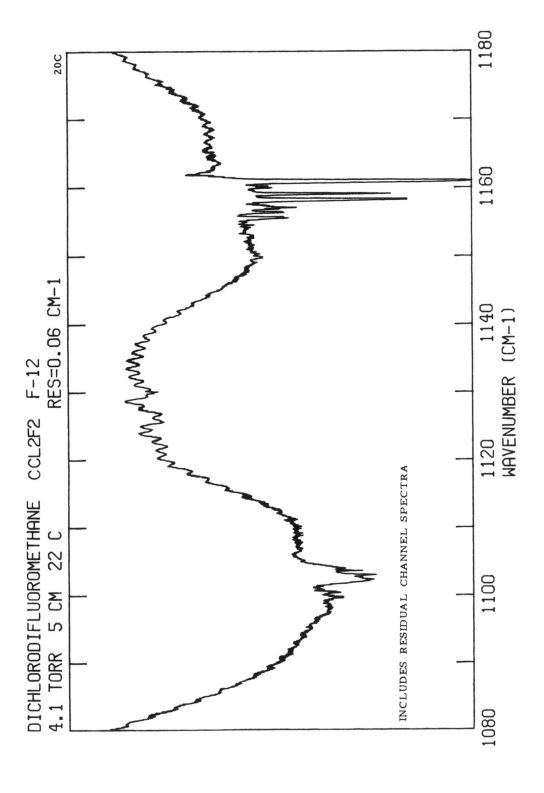

DICHLORODIFLUOROMETHANE CCL2F2 F-12
4.1 TORR 5 CM 22 C RES=0.06 CM-1

20C

INCLUDES RESIDUAL CHANNEL SPECTRA

WAVENUMBER (CM-1)

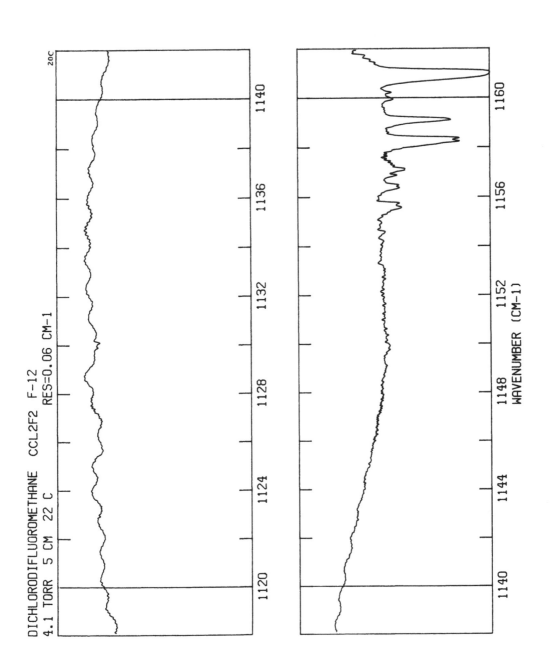

DICHLORODIFLUOROMETHANE CCL2F2 F-12
4.1 TORR 5 CM 22 C   RES=0.06 CM-1

WAVENUMBER (CM-1)

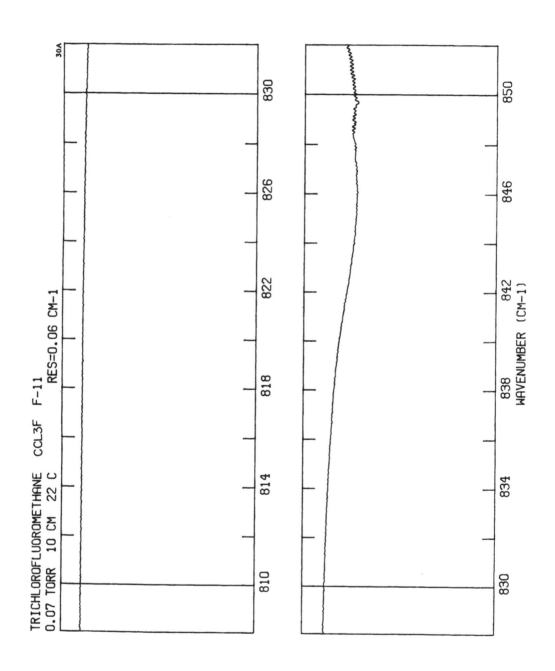

TRICHLOROFLUOROMETHANE   CCL3F   F-11
0.07 TORR   10 CM   22 C   RES=0.06 CM-1

WAVENUMBER (CM-1)

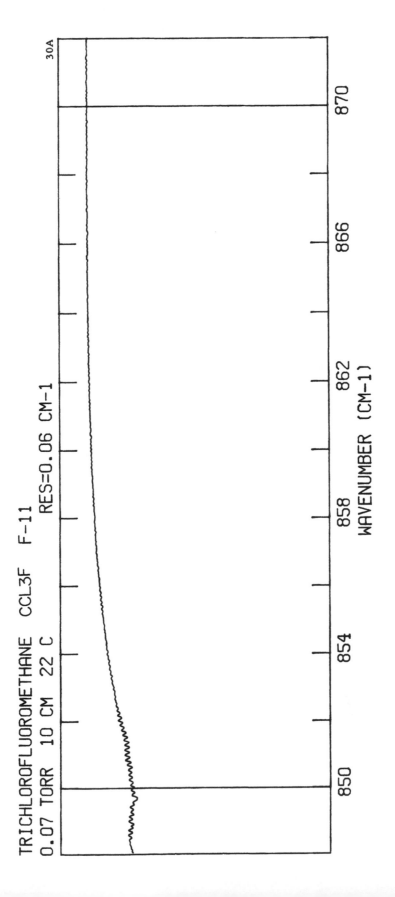

TRICHLOROFLUOROMETHANE   CCL3F   F-11
0.07 TORR   10 CM   22 C   RES=0.06 CM-1

WAVENUMBER (CM-1)

30A

850   854   858   862   866   870

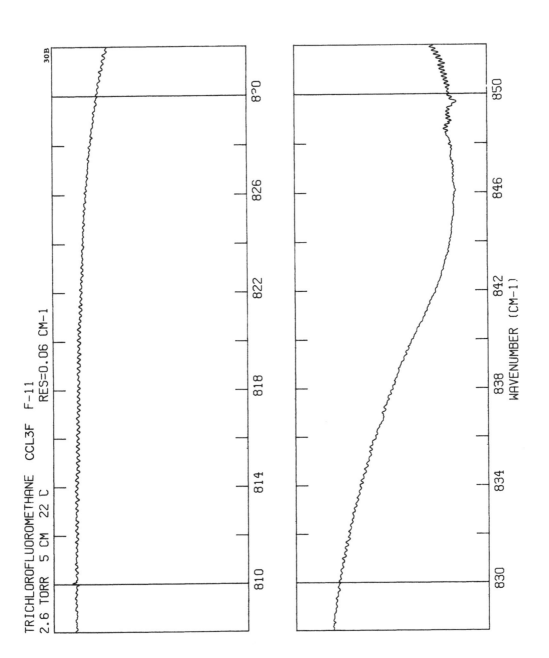

TRICHLOROFLUOROMETHANE CCL3F   F-11
2.6 TORR 5 CM 22 C   RES=0.06 CM-1

30B

WAVENUMBER (CM-1)

41

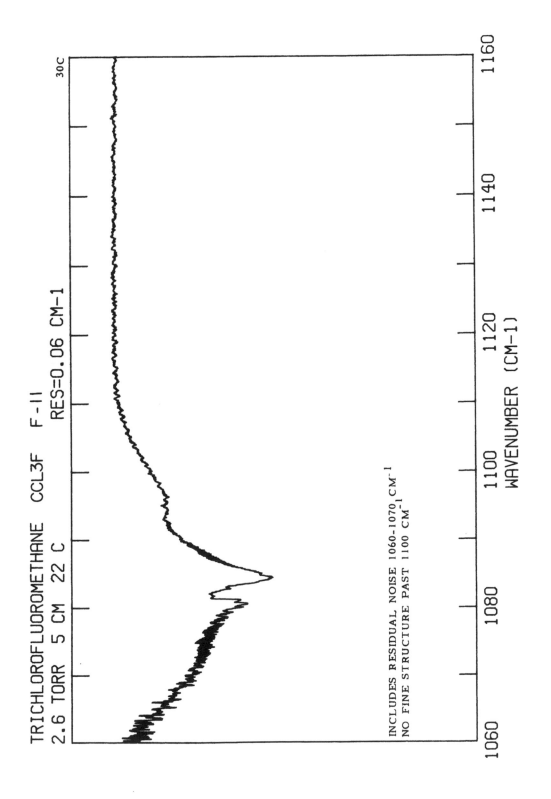

TRICHLOROFLUOROMETHANE CCL3F F-11 RES=0.06 CM-1
2.6 TORR 5 CM 22 C   30C

INCLUDES RESIDUAL NOISE 1060-1070 CM^{-1}
NO FINE STRUCTURE PAST 1100 CM^{-1}

WAVENUMBER (CM-1)

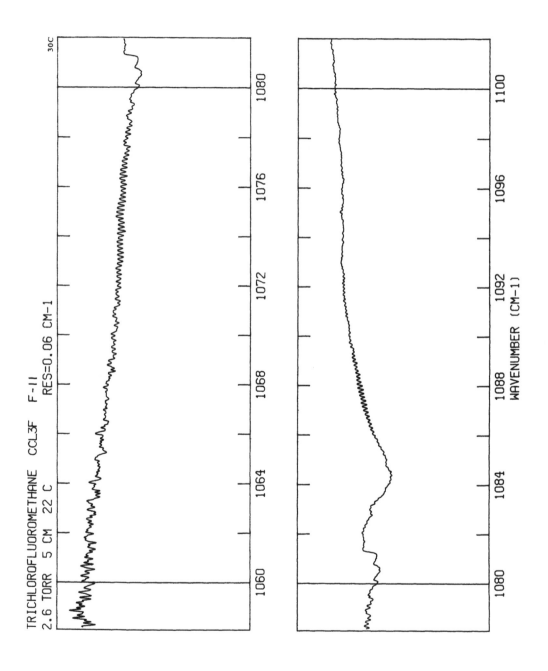

TRICHLOROFLUOROMETHANE   CCL3F   F-11
2.6 TORR   5 CM   22 C   RES=0.06 CM-1
30C

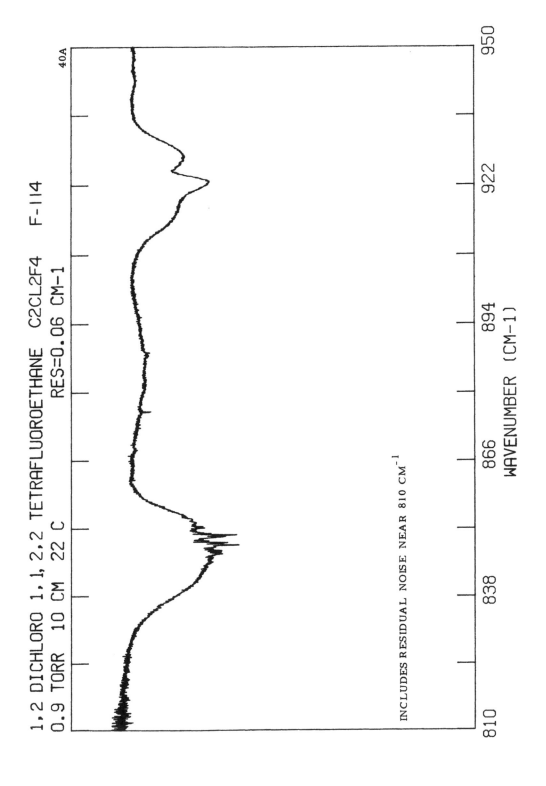

1,2 DICHLORO 1,1,2,2 TETRAFLUOROETHANE C2CL2F4   F-114
0.9 TORR 10 CM 22 C   RES=0.06 CM-1

40A

INCLUDES RESIDUAL NOISE NEAR 810 CM^{-1}

WAVENUMBER (CM-1)

810   838   866   894   922   950

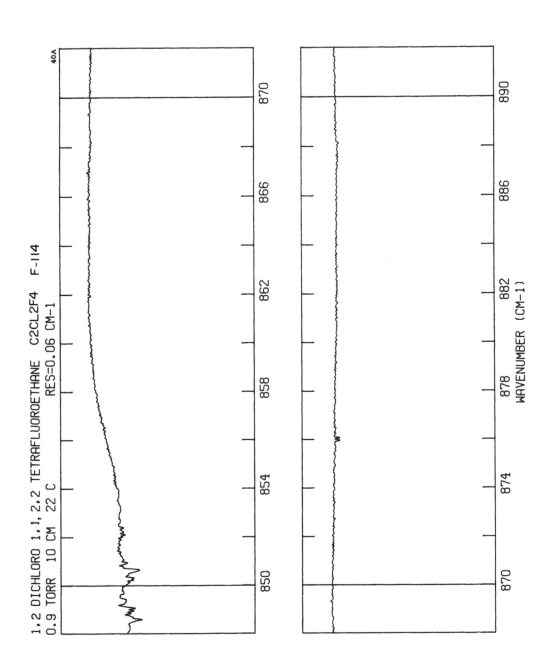

1,2 DICHLORO 1,1,2,2 TETRAFLUOROETHANE C2CL2F4 F-114
0.9 TORR 10 CM 22 C RES=0.06 CM-1

40A

WAVENUMBER (CM-1)

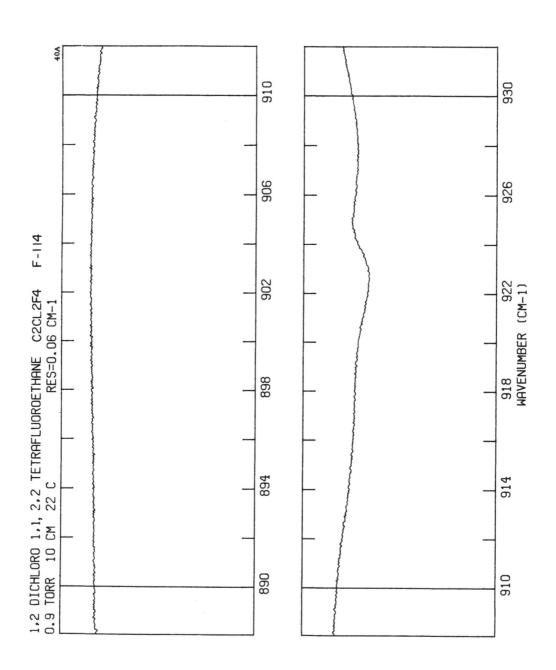

1,2 DICHLORO 1,1,2,2 TETRAFLUOROETHANE   C2CL2F4   F-114
0.9 TORR   10 CM   22 C
RES=0.06 CM-1

1,2 DICHLORO 1,1,2,2 TETRAFLUOROETHANE C2CL2F4 F114
0.9 TORR 10 CM 22 C   RES=0.06 CM-1

40B

NO FINE STRUCTURE RESOLVED
ONLY 1160-1240 CM^{-1} INCLUDED ON EXPANDED PLOTS

WAVENUMBER (CM-1)

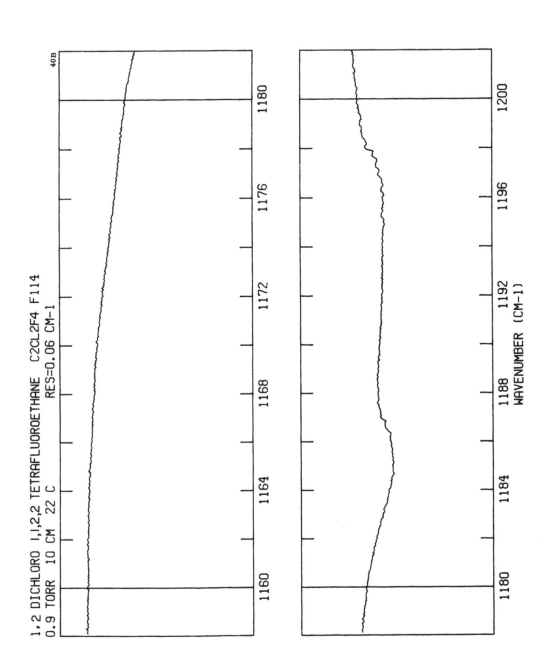

1, 2 DICHLORO 1,1,2,2 TETRAFLUOROETHANE   C2CL2F4 F114
0.9 TORR   10 CM   22 C   RES=0.06 CM-1

40B

WAVENUMBER (CM-1)

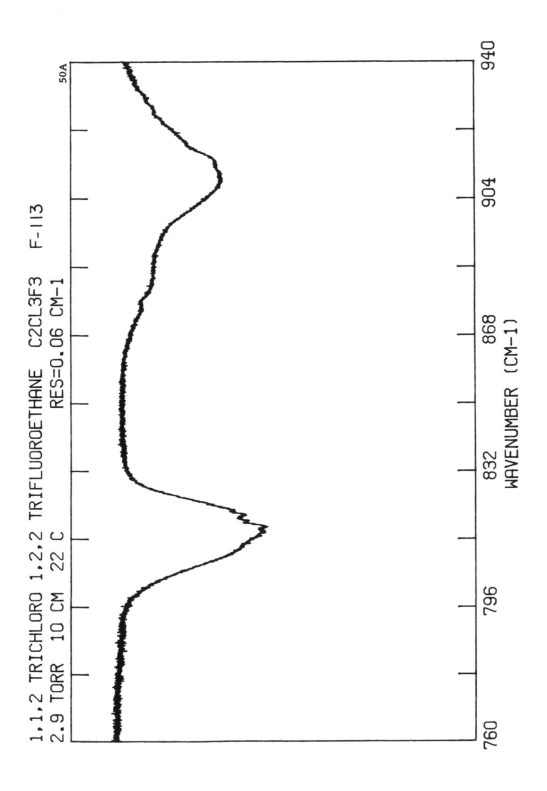

1,1,2 TRICHLORO 1,2,2 TRIFLUOROETHANE C2CL3F3 F-113
2.9 TORR 10 CM 22 C RES=0.06 CM-1

50A

WAVENUMBER (CM-1)

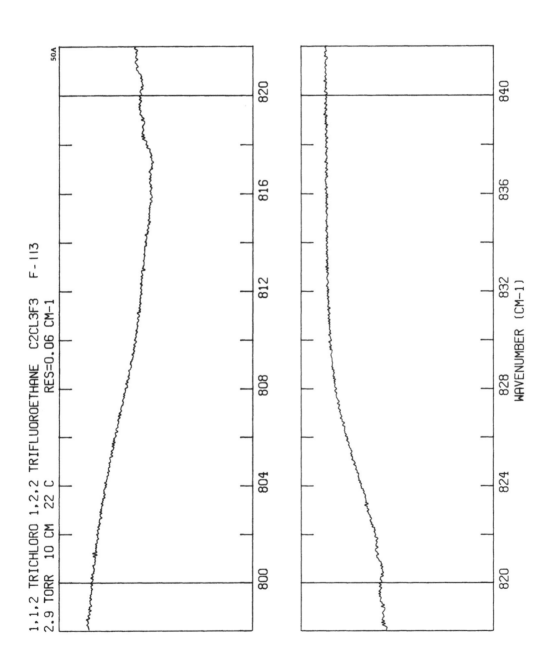

1,1,2 TRICHLORO 1,2,2 TRIFLUOROETHANE C2CL3F3   F-113
2.9 TORR 10 CM 22 C   RES=0.06 CM-1

50A

WAVENUMBER (CM-1)

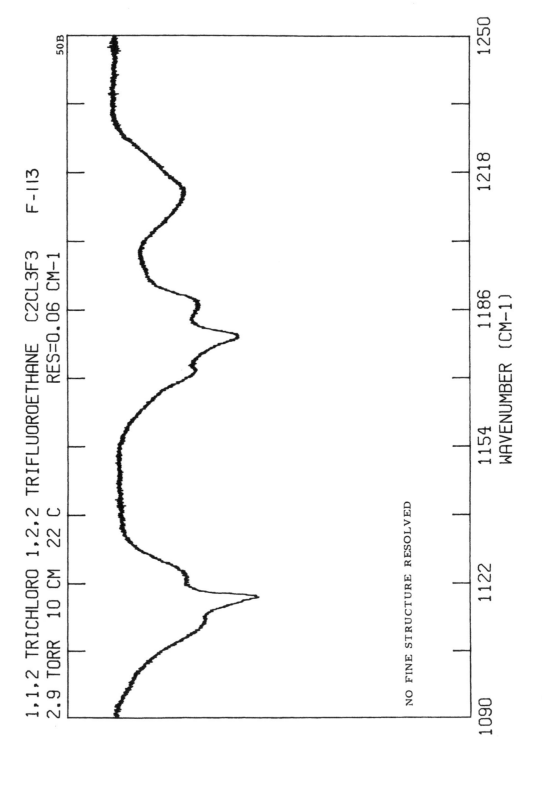

1,1,2 TRICHLORO 1,2,2 TRIFLUOROETHANE   C2CL3F3   F-113
2.9 TORR   10 CM   22 C   RES=0.06 CM-1

50B

NO FINE STRUCTURE RESOLVED

WAVENUMBER (CM-1)

1090   1122   1154   1186   1218   1250

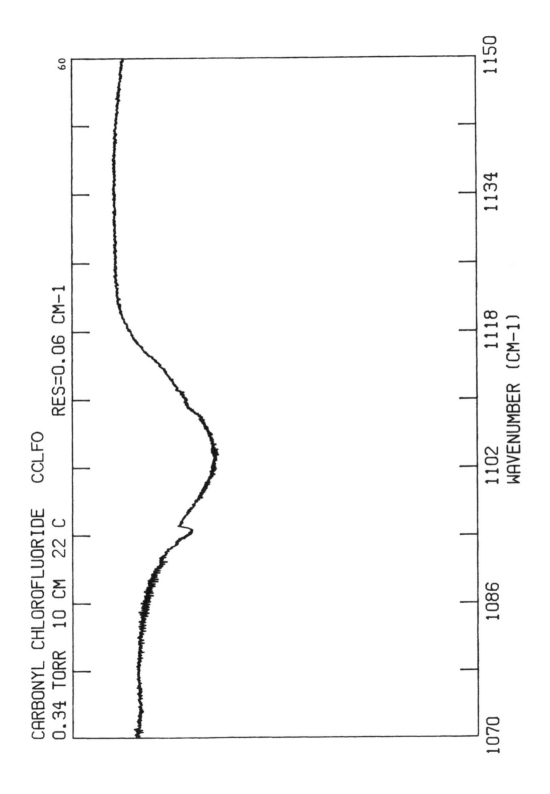

CARBONYL CHLOROFLUORIDE   CCLFO   RES=0.06 CM-1

0.34 TORR   10 CM   22 C

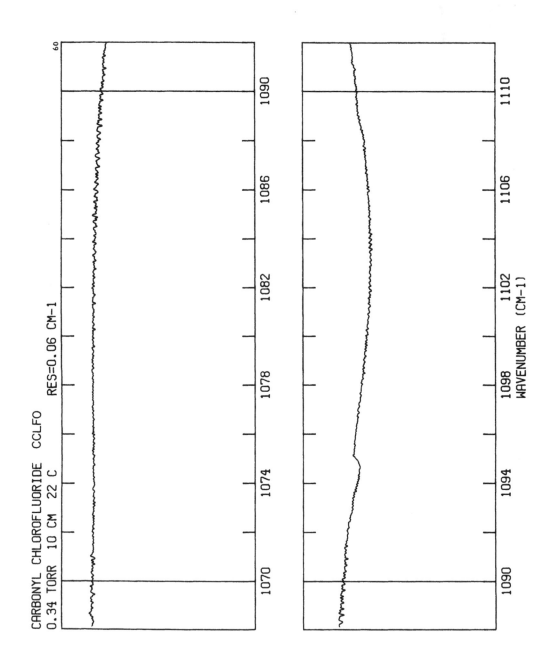

CARBONYL CHLOROFLUORIDE   CCLFO   RES=0.06 CM-1

0.34 TORR   10 CM   22 C

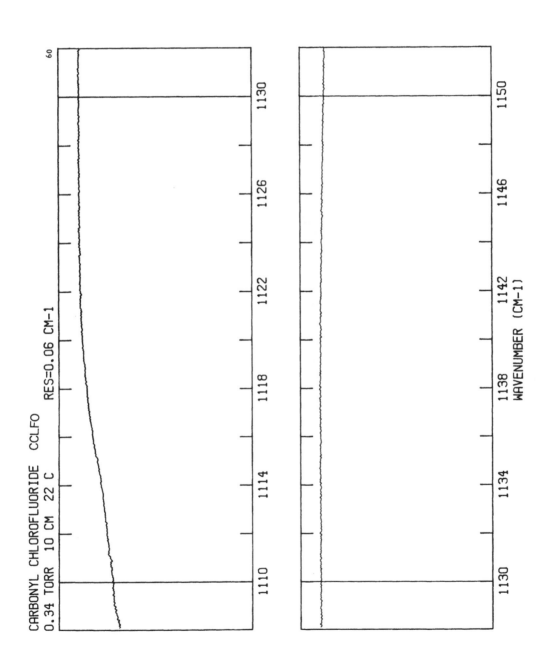

CARBONYL CHLOROFLUORIDE   CCLFO   RES=0.06 CM-1
0.34 TORR   10 CM   22 C

WAVENUMBER (CM-1)

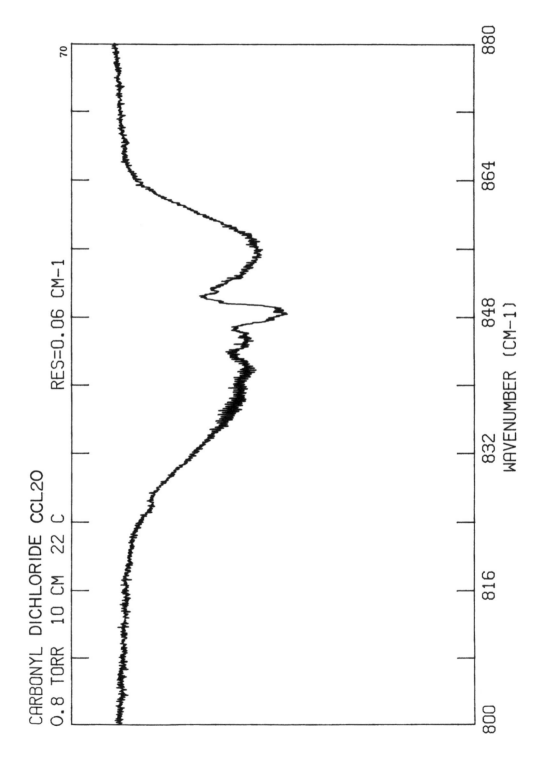

CARBONYL DICHLORIDE CCL2O

0.8 TORR 10 CM 22 C

RES=0.06 CM-1

WAVENUMBER (CM-1)

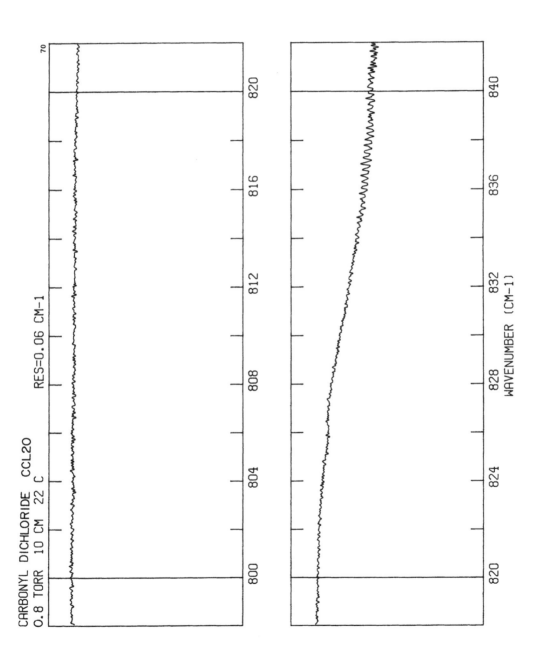

CARBONYL DICHLORIDE CCL2O   RES=0.06 CM-1
0.8 TORR   10 CM   22 C

WAVENUMBER (CM-1)

61

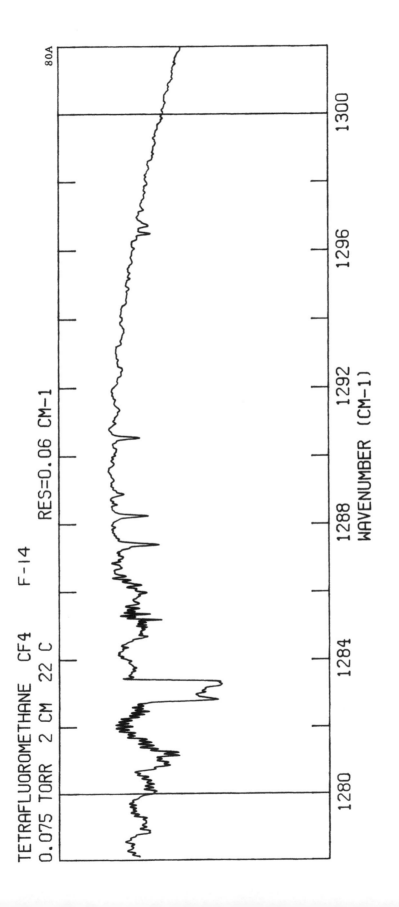

TETRAFLUOROMETHANE   CF4   F-14   RES=0.06 CM-1

0.075 TORR   2 CM   22 C

WAVENUMBER (CM-1)

80A

1300   1296   1292   1288   1284   1280

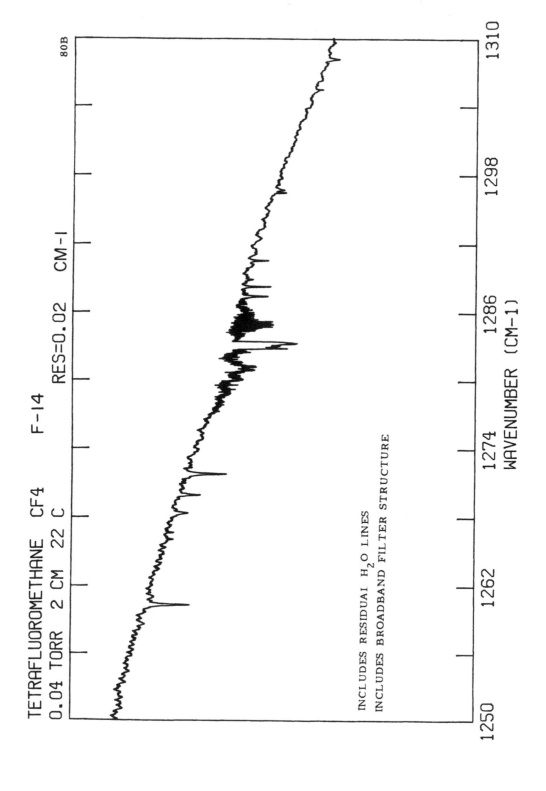

TETRAFLUOROMETHANE CF4   F-14

0.04 TORR   2 CM   22 C   RES=0.02   CM-I

80B

INCLUDES RESIDUAI H₂O LINES

INCLUDES BROADBAND FILTER STRUCTURE

WAVENUMBER (CM-1)

1250   1262   1274   1286   1298   1310

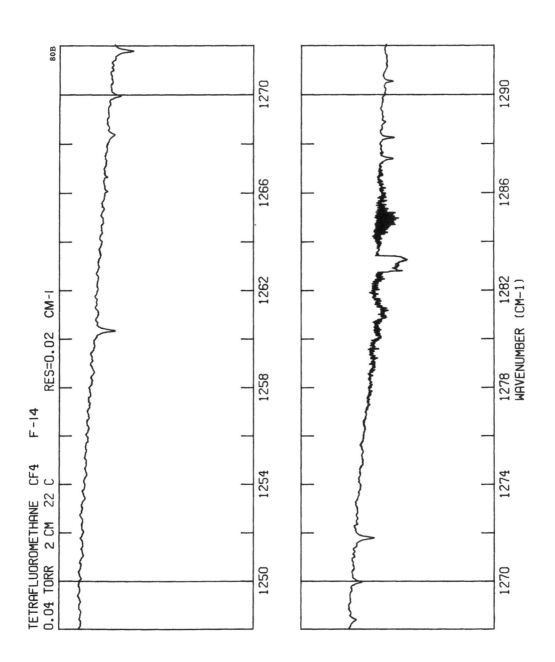

TETRAFLUOROMETHANE CF4   F-14
0.04 TORR   2 CM 22 C   RES=0.02 CM-1

80B

1250   1254   1258   1262   1266   1270

1270   1274   1278   1282   1286   1290
WAVENUMBER (CM-1)

TETRAFLUOROMETHANE, METHANE AND NITROUS OXIDE   CF4+CH4+N2O

22 C   RES=0.02CM-1

P_{CF_4} = .04 torr (2 cm)

P_{CH_4} = 42 torr

P_{N_2O} = 33 torr } (5 cm)

INCLUDES RESIDUAL CHANNEL SPECTRA
INCLUDES BROADBAND FILTER STRUCTURE

90

1050   1110   1170   1230   1290   1350

WAVENUMBER (CM-1)

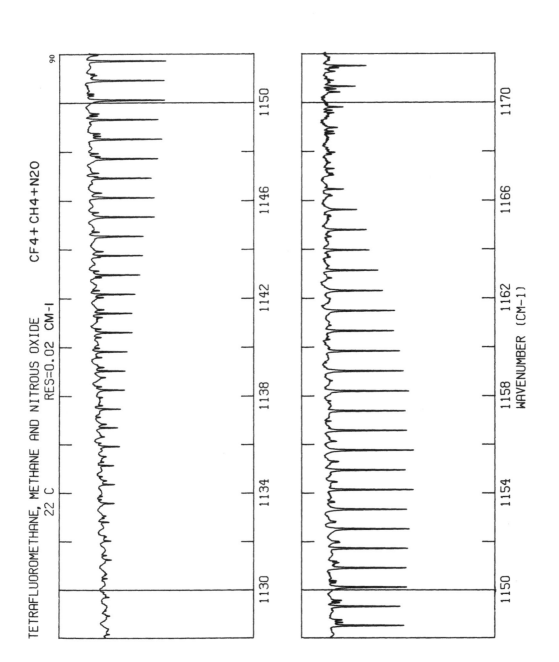

TETRAFLUOROMETHANE, METHANE AND NITROUS OXIDE

22 C

CF4+CH4+N2O

RES=0.02 CM-I

WAVENUMBER (CM-1)

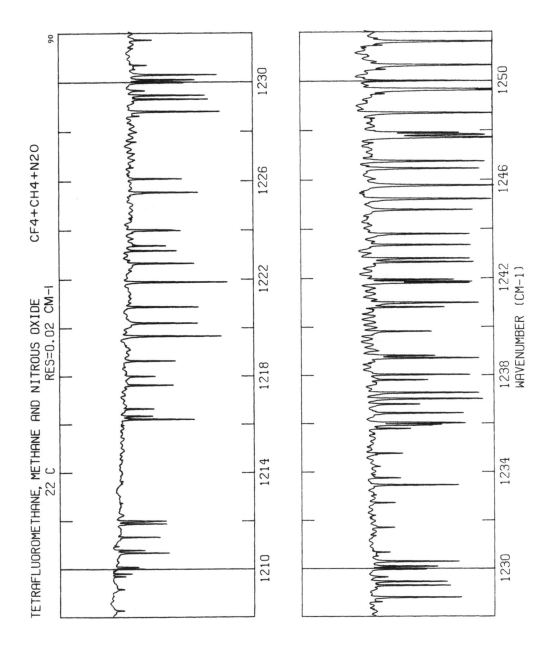

TETRAFLUOROMETHANE, METHANE AND NITROUS OXIDE   CF4+CH4+N2O
22 C   RES=0.02 CM-I

WAVENUMBER (CM-1)

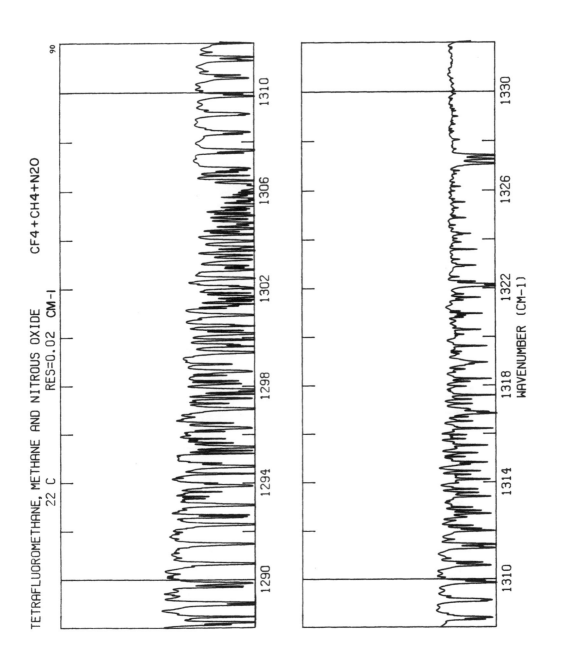

TETRAFLUOROMETHANE, METHANE AND NITROUS OXIDE   CF4+CH4+N2O
22 C   RES=0.02 CM-1

WAVENUMBER (CM-1)

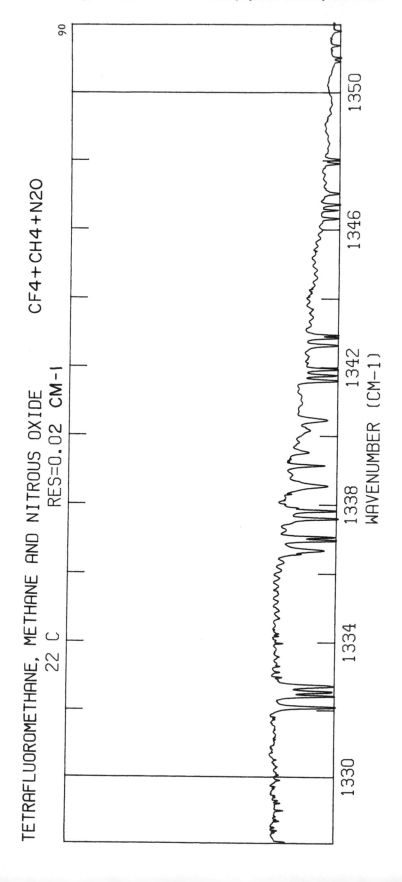

TETRAFLUOROMETHANE, METHANE AND NITROUS OXIDE   CF4+CH4+N2O

22 C   RES=0.02 CM-1

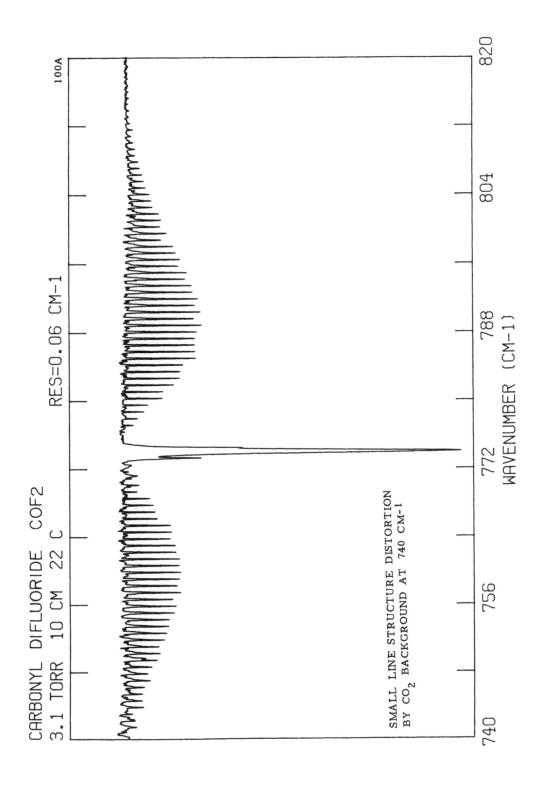

CARBONYL DIFLUORIDE COF2

RES=0.06 CM-1

3.1 TORR 10 CM 22 C

100A

WAVENUMBER (CM-1)

SMALL LINE STRUCTURE DISTORTION
BY CO$_2$ BACKGROUND AT 740 CM-1

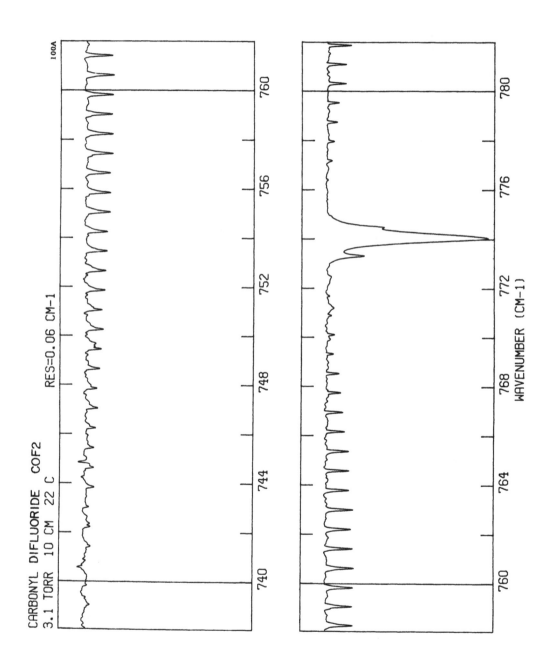

CARBONYL DIFLUORIDE COF2   RES=0.06 CM-1
3.1 TORR 10 CM 22 C

WAVENUMBER (CM-1)

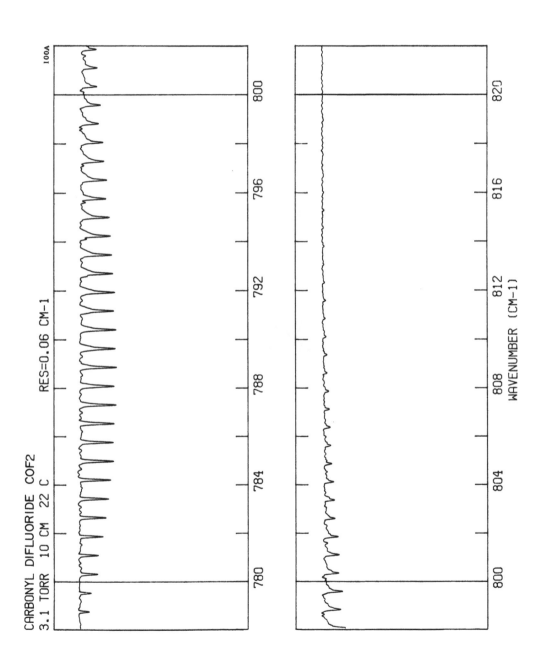

CARBONYL DIFLUORIDE COF2

3.1 TORR   10 CM   22 C

RES=0.06 CM-1

100A

WAVENUMBER (CM-1)

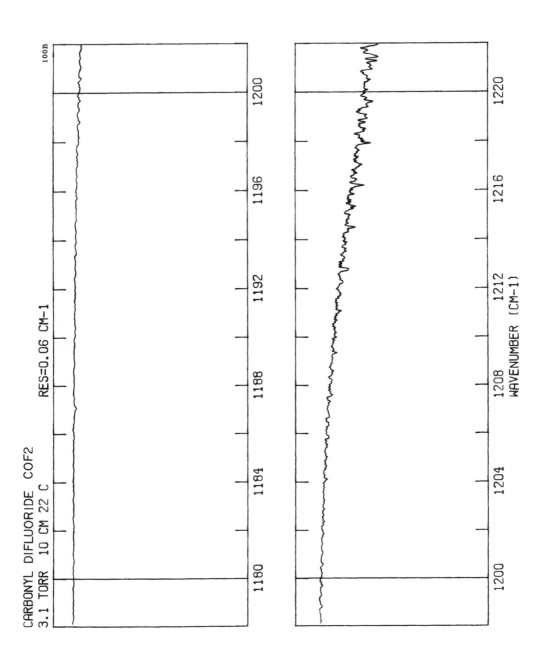

CARBONYL DIFLUORIDE COF2   RES=0.06 CM-1
3.1 TORR 10 CM 22 C

100B

1180   1184   1188   1192   1196   1200

1200   1204   1208   1212   1216   1220
                    WAVENUMBER (CM-1)

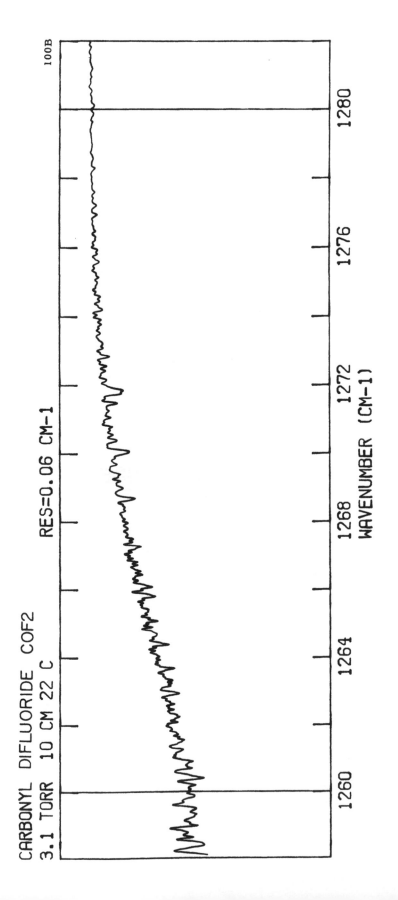

CARBONYL DIFLUORIDE COF2

3.1 TORR 10 CM 22 C

RES=0.06 CM-1

WAVENUMBER (CM-1)

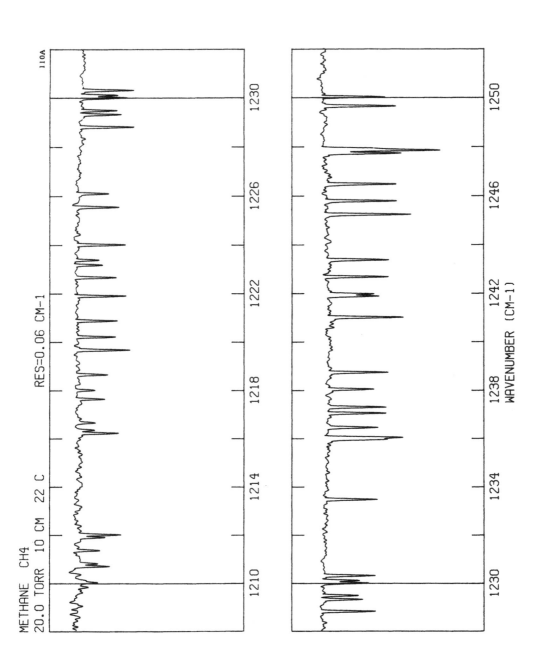

METHANE CH4
20.0 TORR 10 CM 22 C   RES=0.06 CM-1

110A

WAVENUMBER (CM-1)

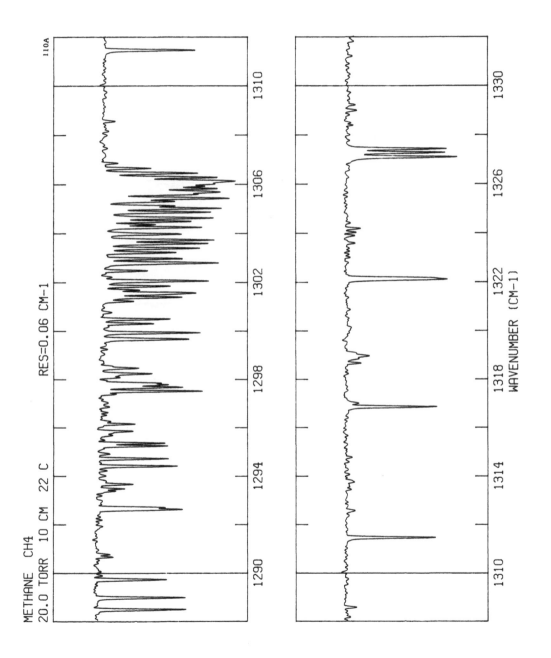

METHANE CH4   RES=0.06 CM-1
20.0 TORR 10 CM 22 C

110A

1290 1294 1298 1302 1306 1310

1310 1314 1318 1322 1326 1330
WAVENUMBER (CM-1)

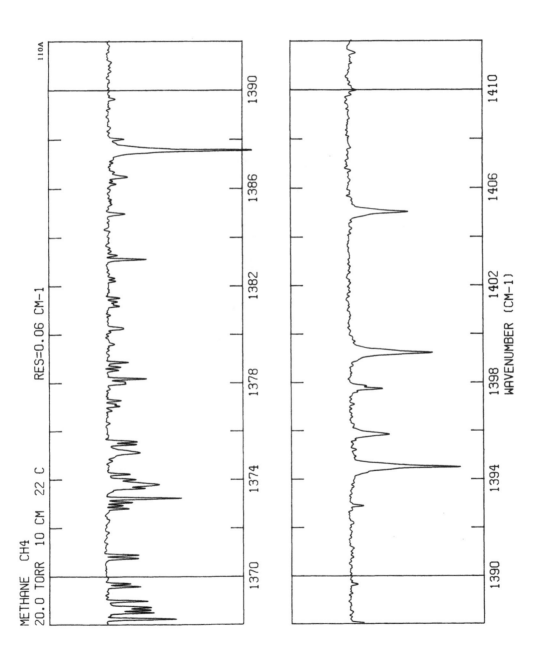

METHANE CH4
20.0 TORR 10 CM 22 C
RES=0.06 CM-1

110A

WAVENUMBER (CM-1)

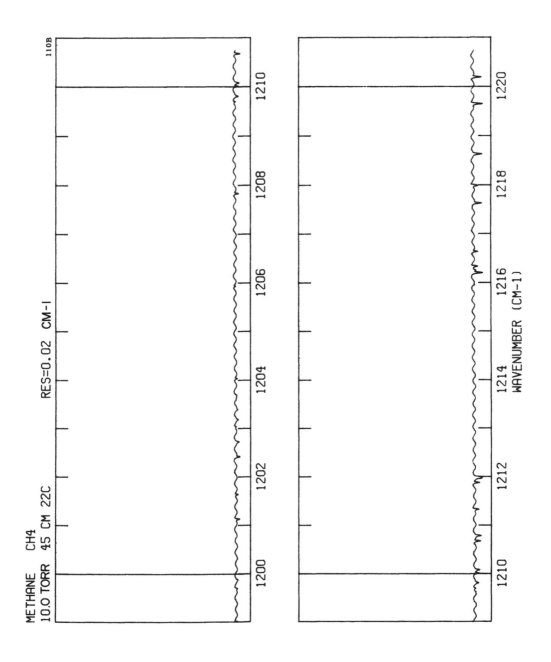

METHANE CH4   RES=0.02 CM-1
10.0 TORR 45 CM 22C

110B

WAVENUMBER (CM-1)

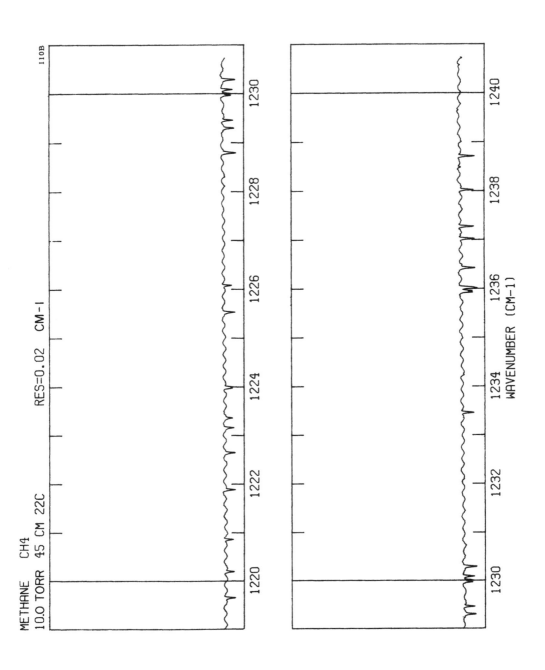

METHANE   CH4
10.0 TORR 45 CM 22C   RES=0.02   CM-1

110B

WAVENUMBER (CM-1)

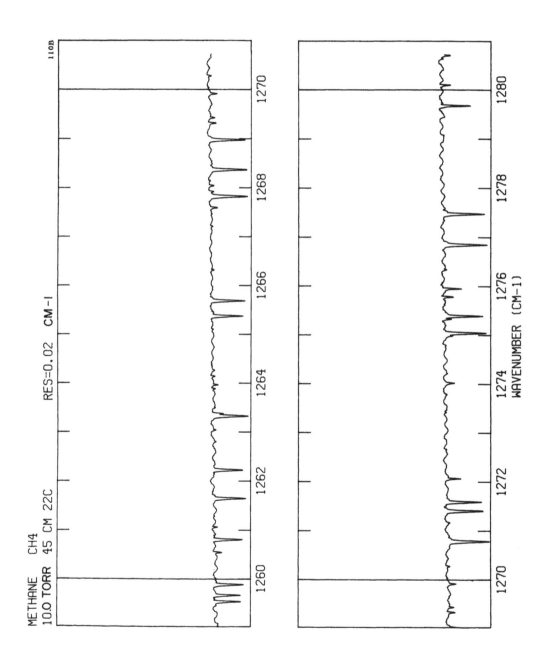

METHANE   CH4
10.0 TORR   45 CM 22C   RES=0.02   CM-I   110B

WAVENUMBER (CM-1)

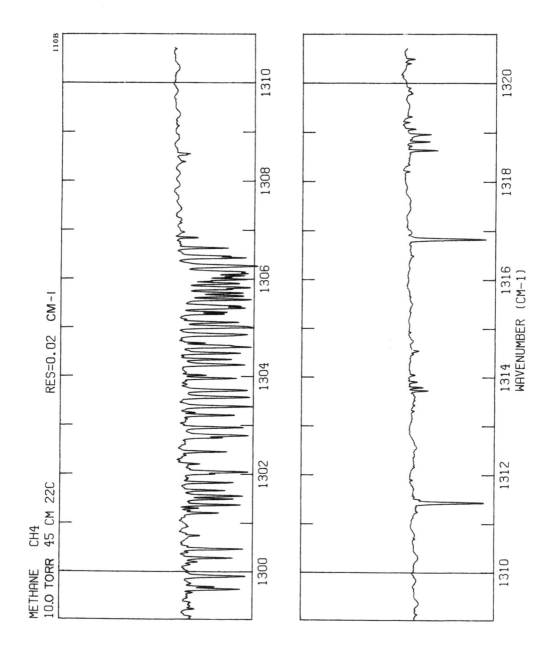

METHANE   CH4
10.0 TORR   45 CM 22C   RES=0.02   CM-1   110B

WAVENUMBER (CM-1)

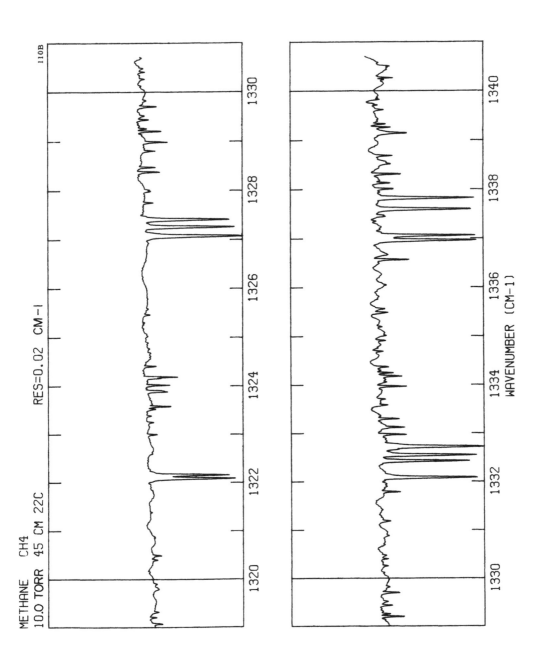

METHANE   CH4
10.0 TORR   45 CM   22C

RES=0.02   CM-1

110B

1320   1322   1324   1326   1328   1330

1330   1332   1334   1336   1338   1340

WAVENUMBER (CM-1)

95

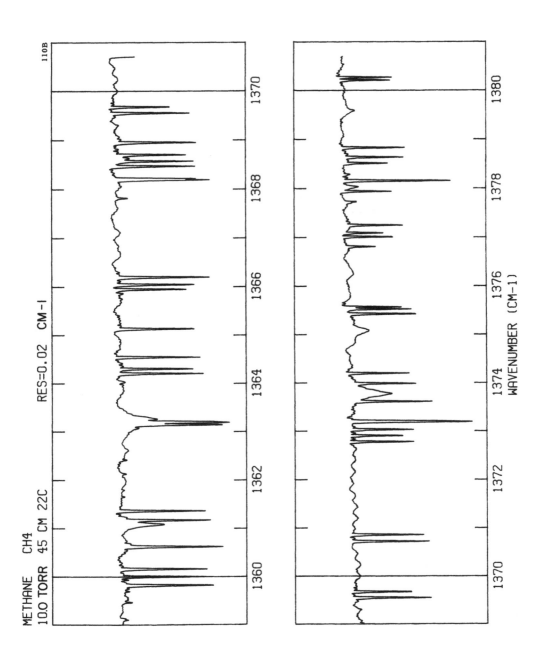

METHANE   CH4
10.0 TORR   45 CM 22C   RES=0.02   CM-I   110B

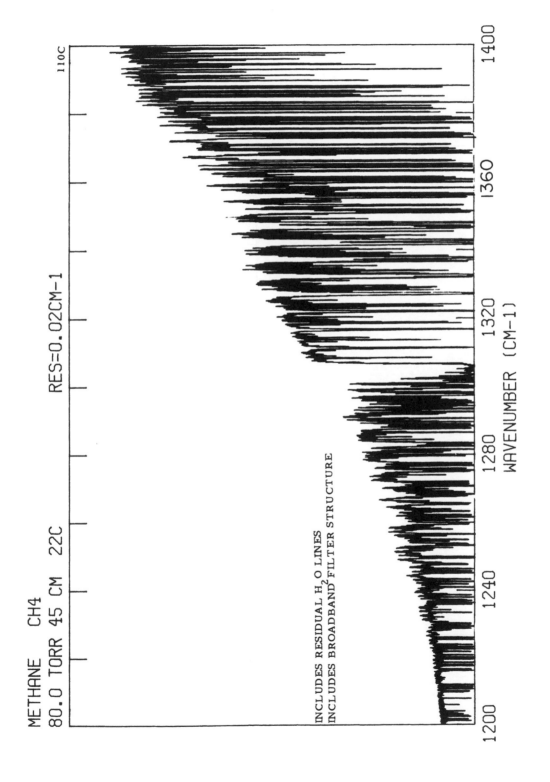

METHANE   CH4   RES=0.02CM-1
80.0 TORR 45 CM 22C

110C

INCLUDES RESIDUAL H$_2$O LINES
INCLUDES BROADBAND FILTER STRUCTURE

WAVENUMBER (CM-1)

1200   1240   1280   1320   1360   1400

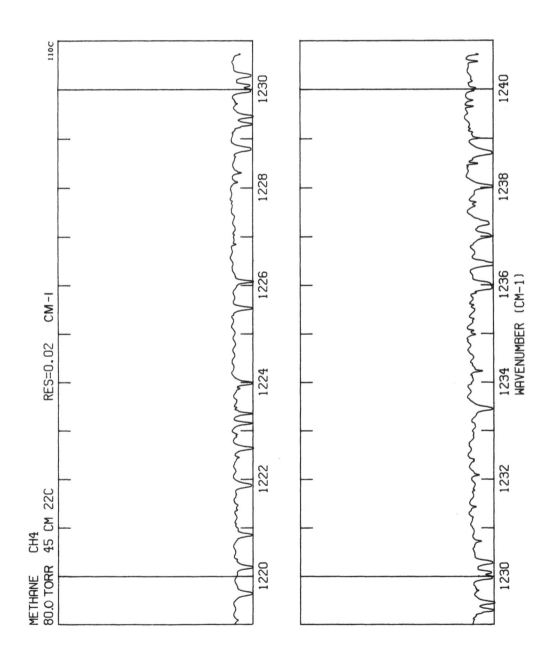

METHANE   CH4
80.0 TORR   45 CM   22C   RES=0.02   CM-I   110C

WAVENUMBER (CM-1)

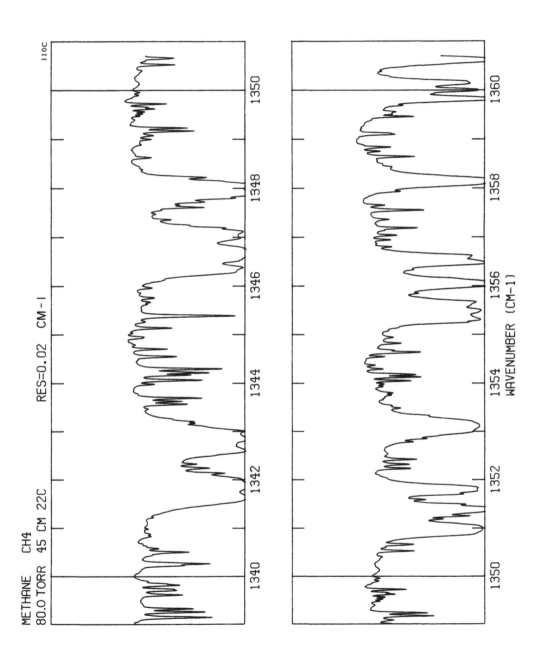

METHANE   CH4
80.0 TORR   45 CM 22C   RES=0.02 CM-1

WAVENUMBER (CM-1)

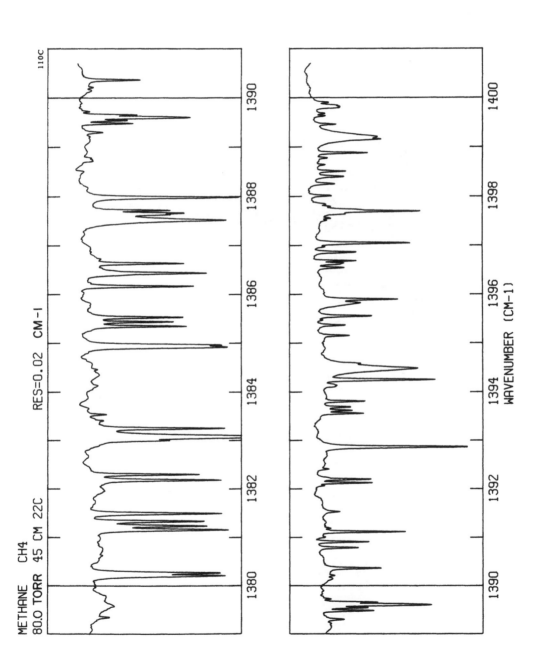

METHANE   CH4
80.0 TORR   45 CM   22C
RES=0.02   CM-I

WAVENUMBER (CM-1)

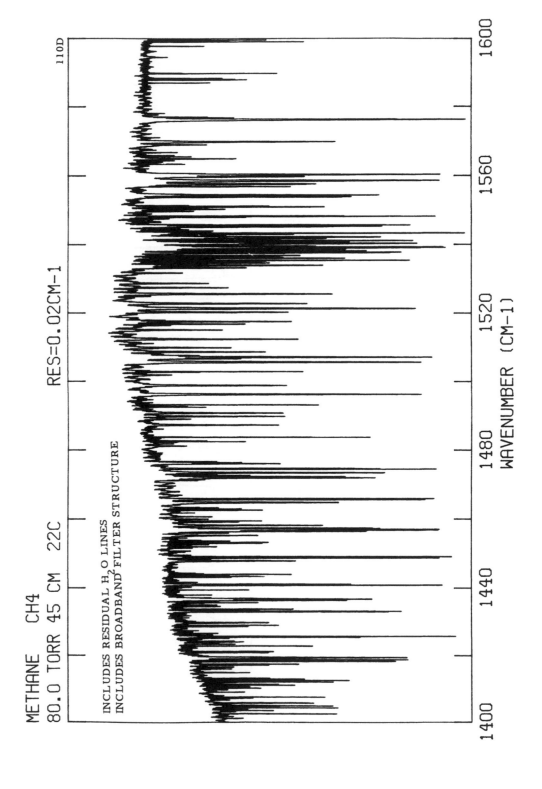

METHANE   CH4

80.0 TORR 45 CM   22C

RES=0.02CM-1

110D

INCLUDES RESIDUAL H₂O LINES
INCLUDES BROADBAND FILTER STRUCTURE

WAVENUMBER (CM-1)

1400   1440   1480   1520   1560   1600

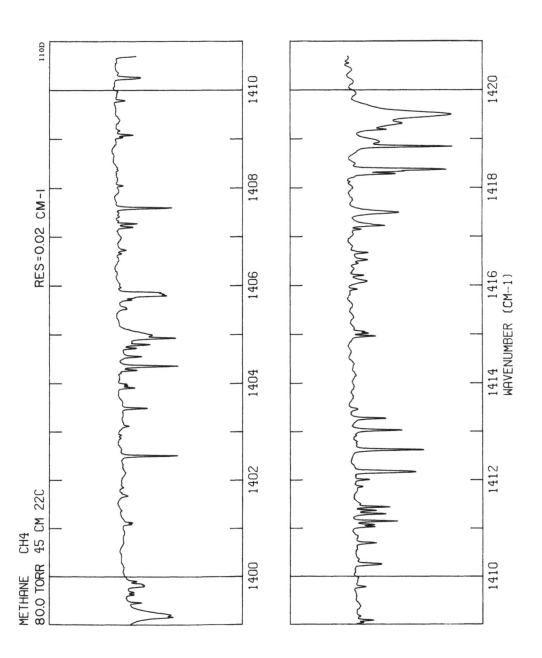

METHANE   CH4   RES=0.02 CM-1   110D
80.0 TORR 45 CM 22C

WAVENUMBER (CM-1)

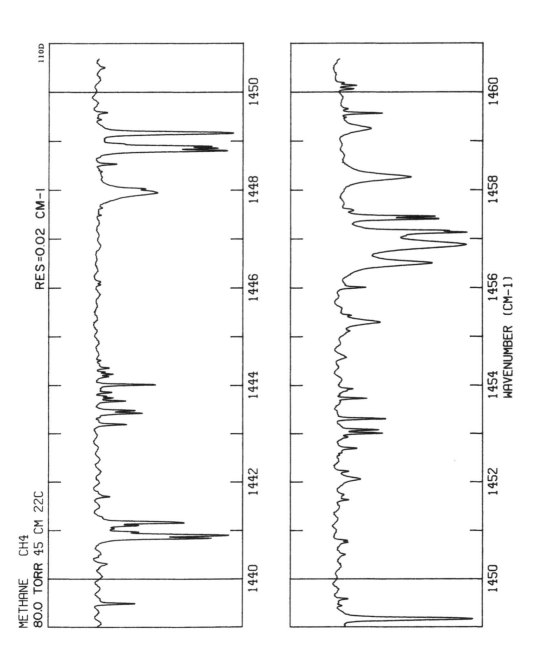

METHANE CH4
80.0 TORR 45 CM 22C

RES=0.02 CM-1

110D

WAVENUMBER (CM-1)

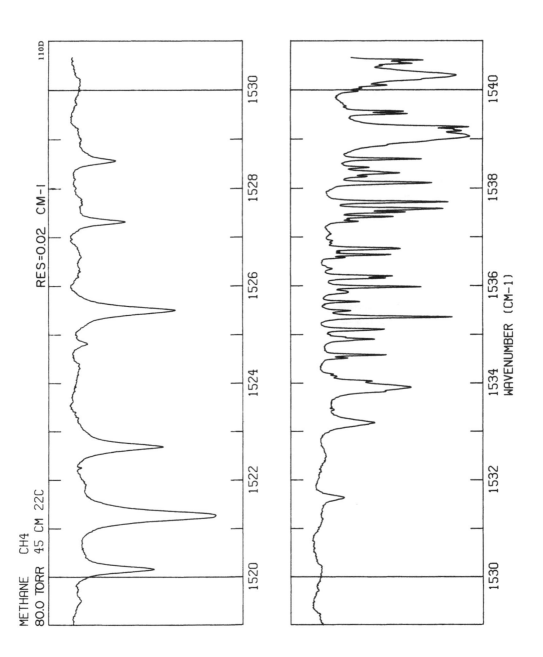

METHANE CH4
80.0 TORR 45 CM 22C

RES=0.02 CM-1

110D

WAVENUMBER (CM-1)

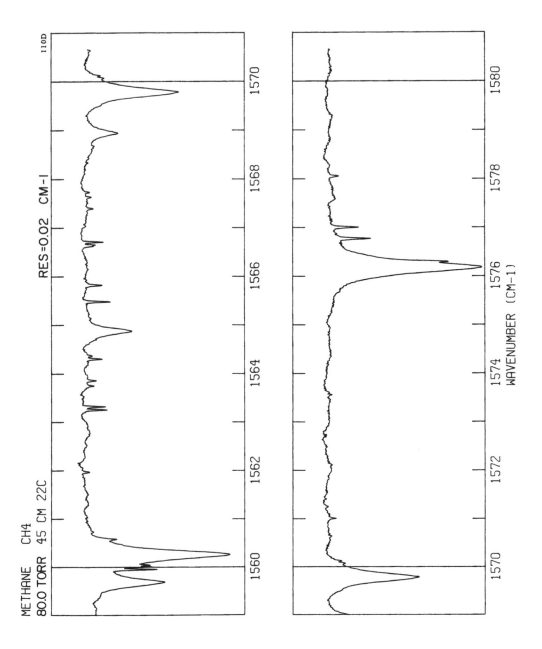

METHANE CH4
80.0 TORR 45 CM 22C

RES=0.02 CM-I

110D

WAVENUMBER (CM-1)

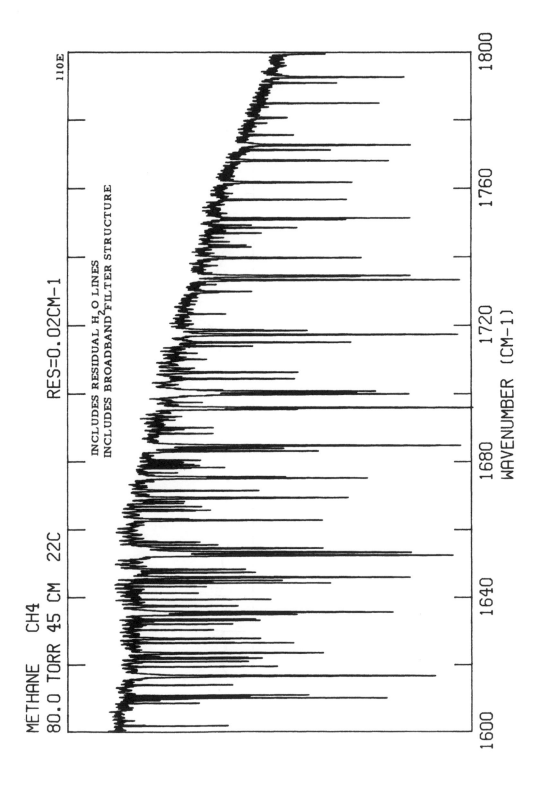

METHANE CH4

80.0 TORR 45 CM 22C

RES=0.02CM−1

110E

INCLUDES RESIDUAL H$_2$O LINES
INCLUDES BROADBAND FILTER STRUCTURE

WAVENUMBER (CM−1)

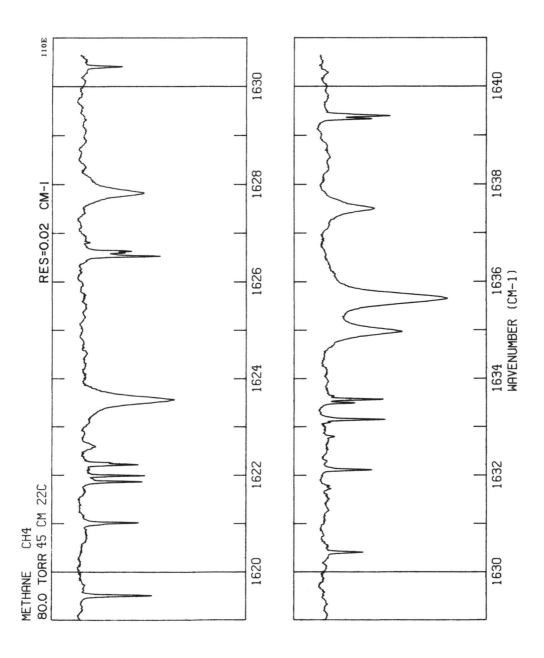

METHANE CH4
80.0 TORR 45 CM 22C

RES=0.02 CM-1

110E

WAVENUMBER (CM-1)

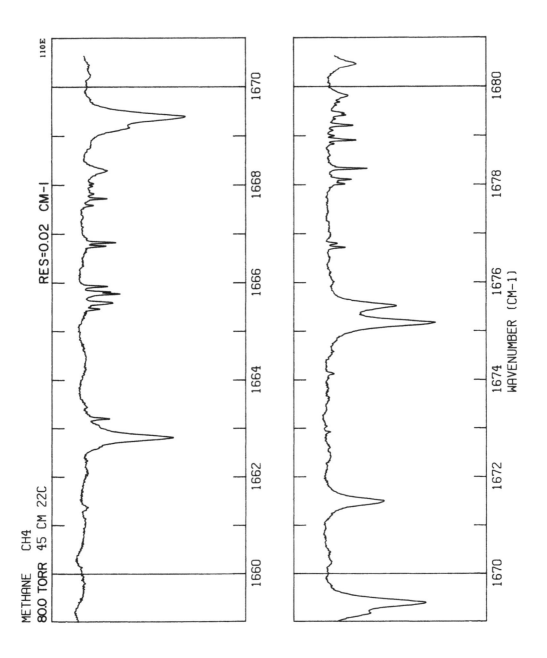

METHANE   CH4
80.0 TORR   45 CM   22C   RES=0.02 CM-1   110E

WAVENUMBER (CM-1)

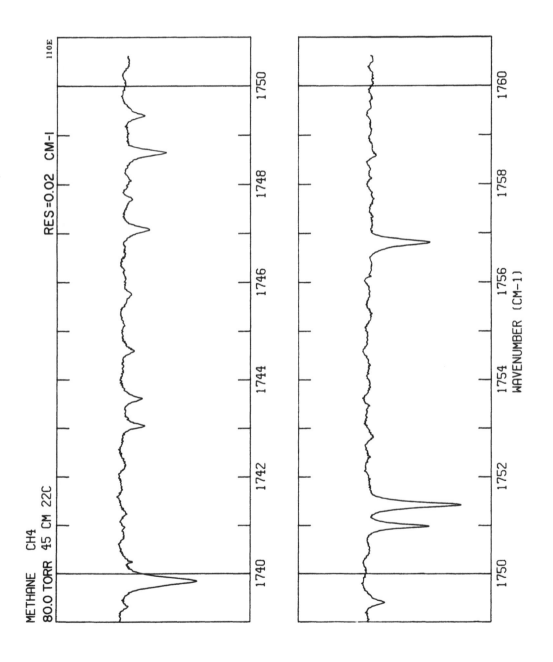

METHANE   CH4
80.0 TORR   45 CM   22C

RES=0.02   CM-1

110E

WAVENUMBER (CM-1)

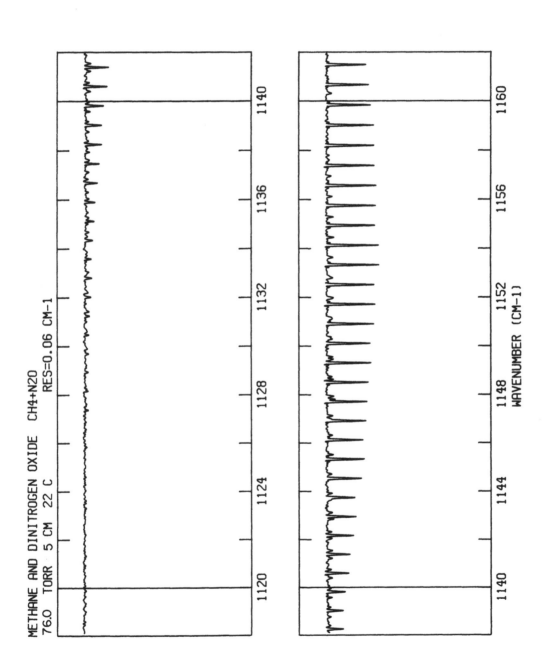

METHANE AND DINITROGEN OXIDE   CH4+N2O
76.0 TORR   5 CM   22 C   RES=0.06 CM-1

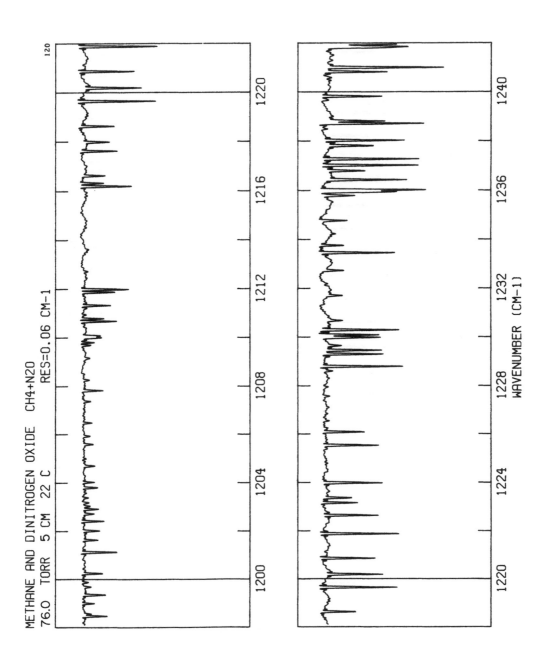

METHANE AND DINITROGEN OXIDE CH4+N20
76.0 TORR 5 CM 22 C   RES=0.06 CM-1

WAVENUMBER (CM-1)

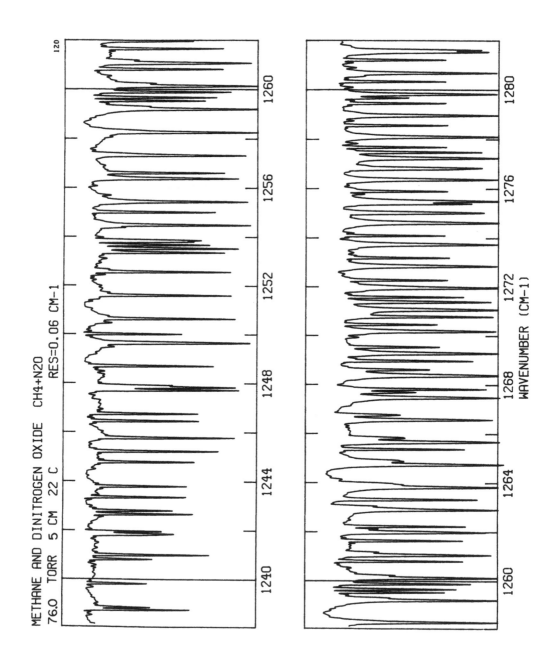

METHANE AND DINITROGEN OXIDE   CH4+N2O
76.0 TORR   5 CM   22 C   RES=0.06 CM-1

WAVENUMBER (CM-1)

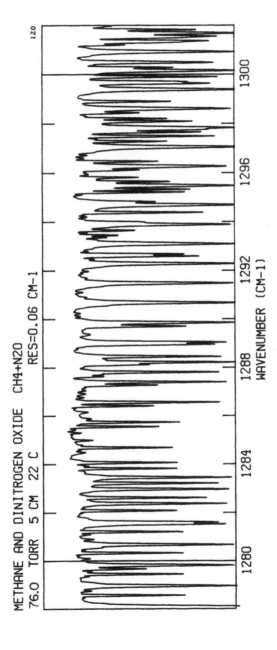

METHANE AND DINITROGEN OXIDE   CH4+N2O
76.0 TORR   5 CM   22 C   RES=0.06 CM-1

WAVENUMBER (CM-1)

1280   1284   1288   1292   1296   1300

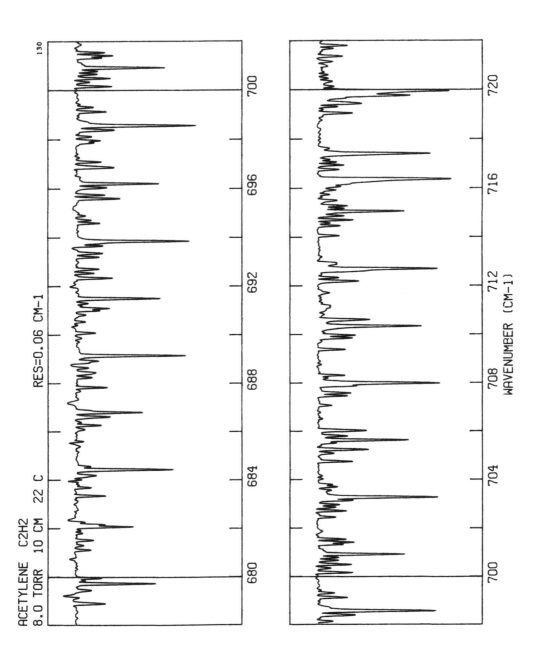

ACETYLENE C2H2
8.0 TORR 10 CM 22 C   RES=0.06 CM-1

WAVENUMBER (CM-1)

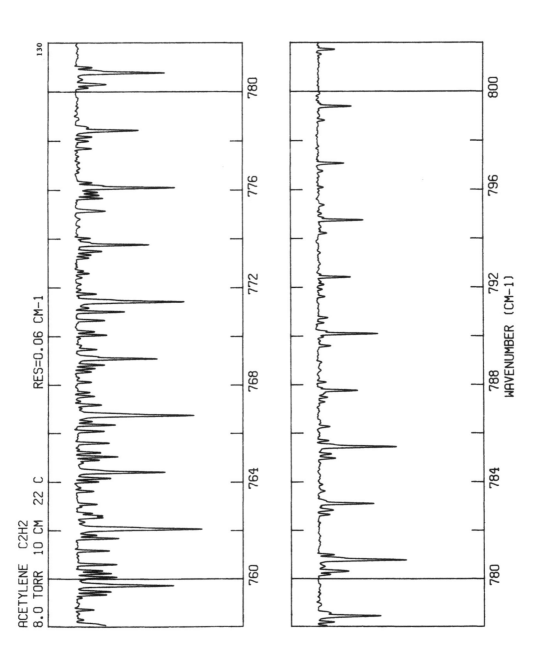

ACETYLENE C2H2   RES=0.06 CM-1
8.0 TORR   10 CM   22 C

WAVENUMBER (CM-1)

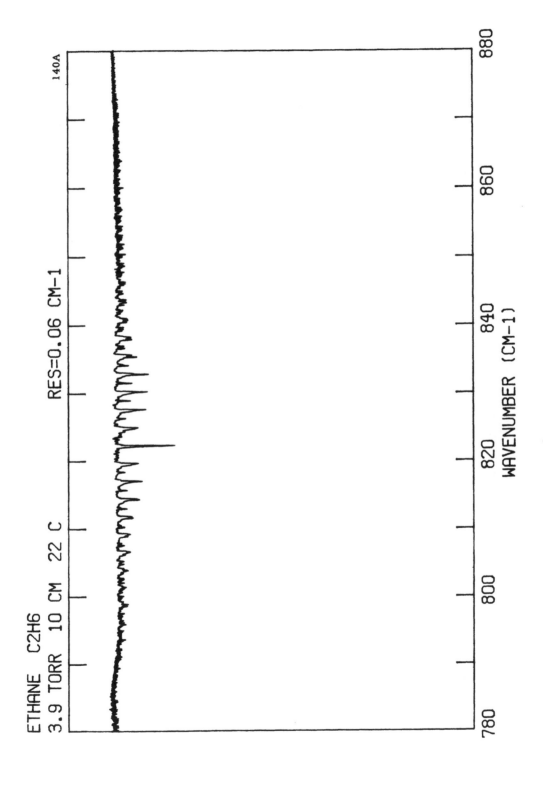

ETHANE C2H6

3.9 TORR 10 CM 22 C

RES=0.06 CM-1

140A

WAVENUMBER (CM-1)

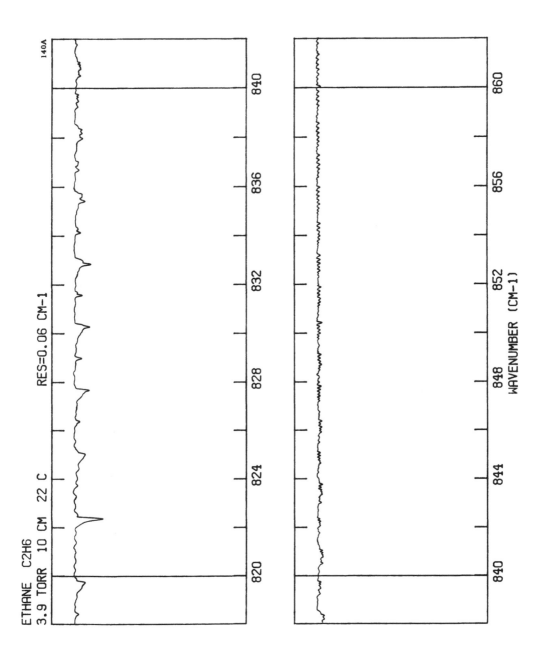

ETHANE C2H6
3.9 TORR 10 CM 22 C
RES=0.06 CM-1
140A

WAVENUMBER (CM-1)

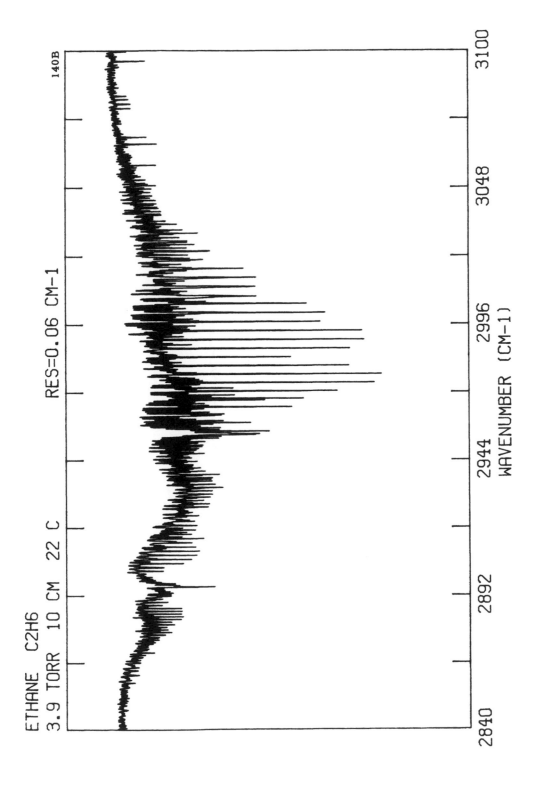

ETHANE   C2H6
3.9 TORR 10 CM 22 C   RES=0.06 CM-1   140B

WAVENUMBER (CM-1)

2840   2892   2944   2996   3048   3100

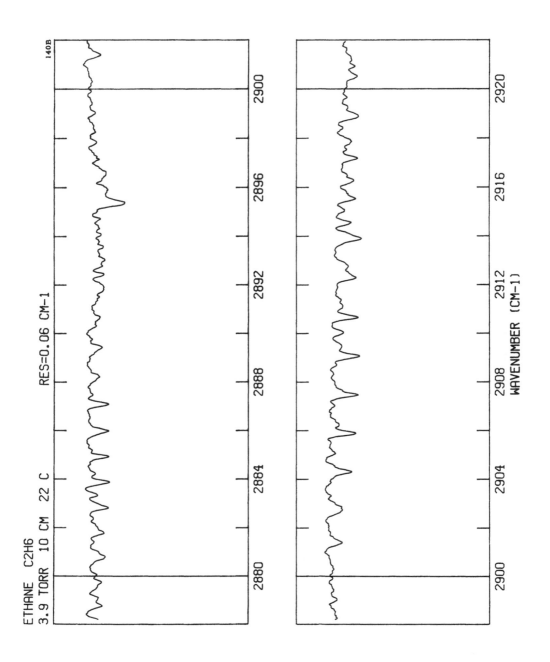

ETHANE C2H6
3.9 TORR 10 CM 22 C   RES=0.06 CM-1

140B

WAVENUMBER (CM-1)

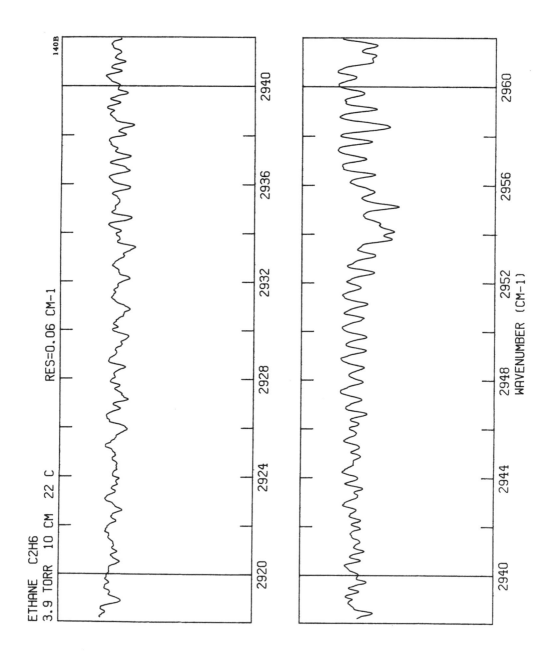

ETHANE   C2H6   RES=0.06 CM-1
3.9 TORR   10 CM   22 C

140B

WAVENUMBER (CM-1)

151

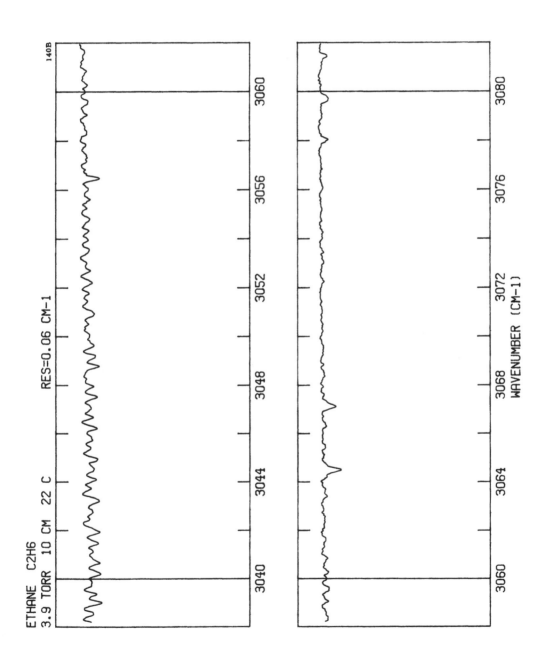

ETHANE C2H6   RES=0.06 CM-1
3.9 TORR 10 CM 22 C

140B

3040   3044   3048   3052   3056   3060

3060   3064   3068   3072   3076   3080
WAVENUMBER (CM-1)

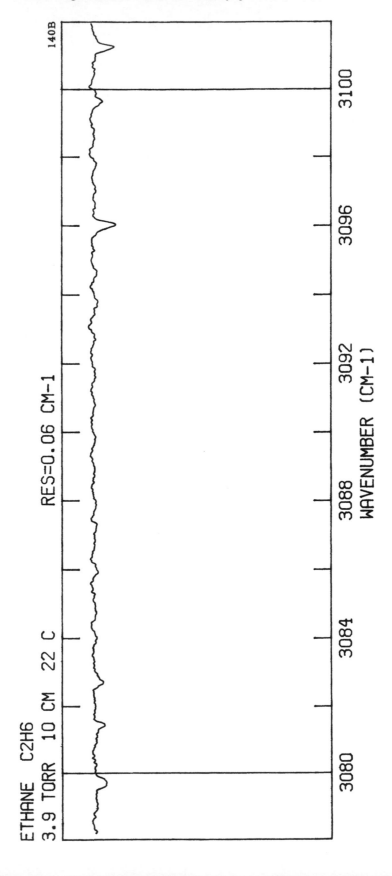

ETHANE C2H6
3.9 TORR 10 CM 22 C
RES=0.06 CM-1

140B

WAVENUMBER (CM-1)

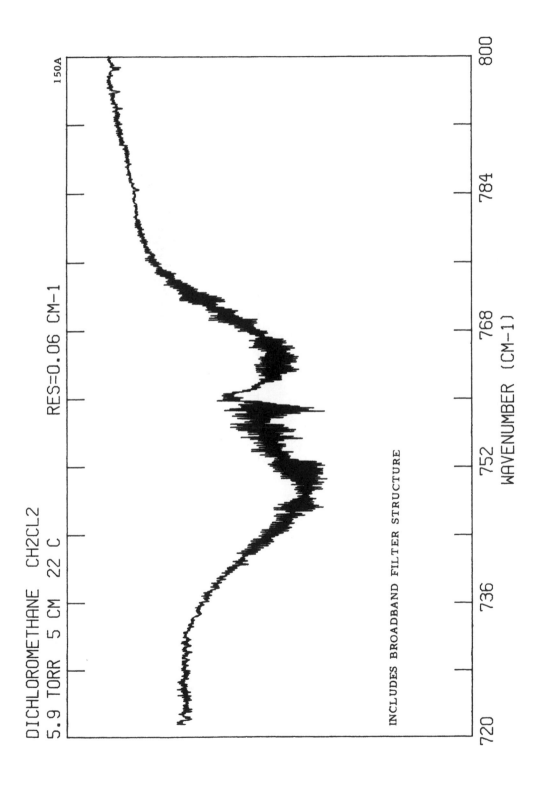

DICHLOROMETHANE CH2CL2
5.9 TORR 5 CM 22 C

RES=0.06 CM-1

150A

INCLUDES BROADBAND FILTER STRUCTURE

WAVENUMBER (CM-1)

720   736   752   768   784   800

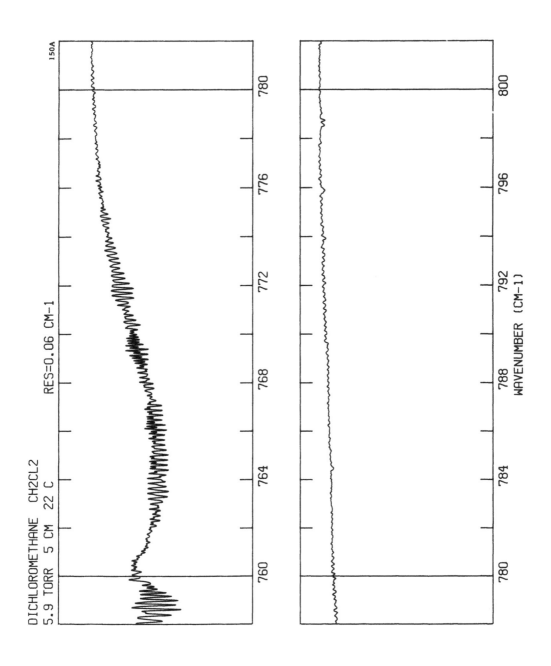

DICHLOROMETHANE CH2CL2   RES=0.06 CM-1
5.9 TORR 5 CM 22 C

150A

WAVENUMBER (CM-1)

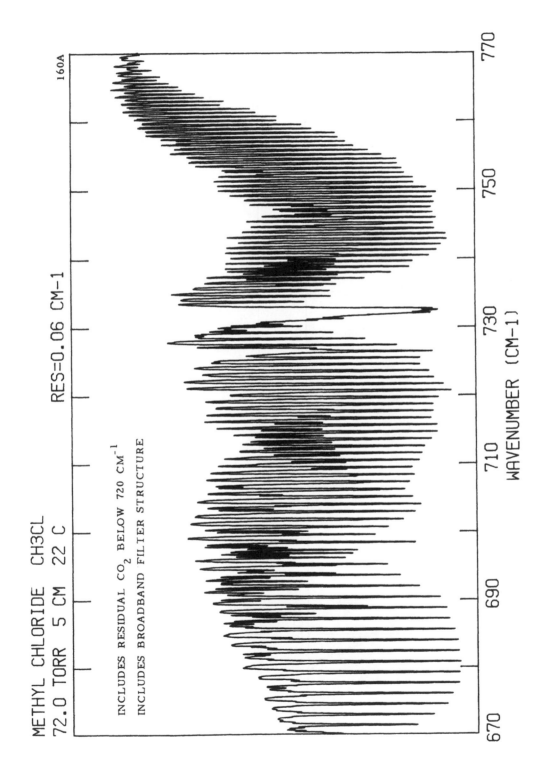

METHYL CHLORIDE CH3CL RES=0.06 CM-1
72.0 TORR 5 CM 22 C

INCLUDES RESIDUAL CO_2 BELOW 720 CM^{-1}
INCLUDES BROADBAND FILTER STRUCTURE

160A

WAVENUMBER (CM-1)

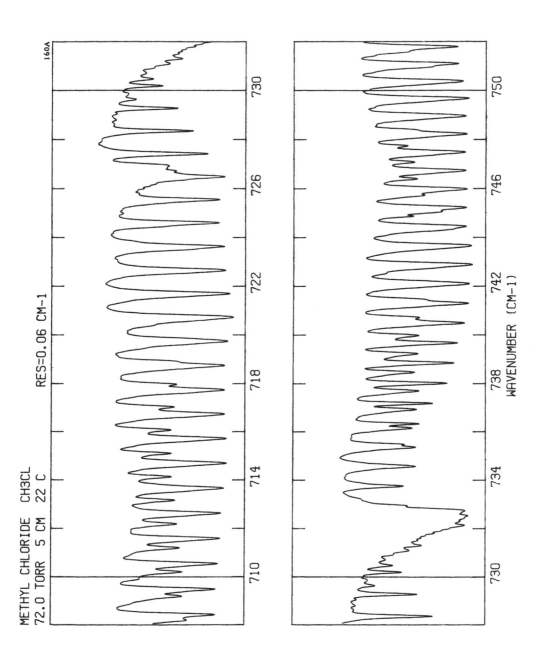

METHYL CHLORIDE CH3CL   RES=0.06 CM-1

72.0 TORR 5 CM 22 C

160A

WAVENUMBER (CM-1)

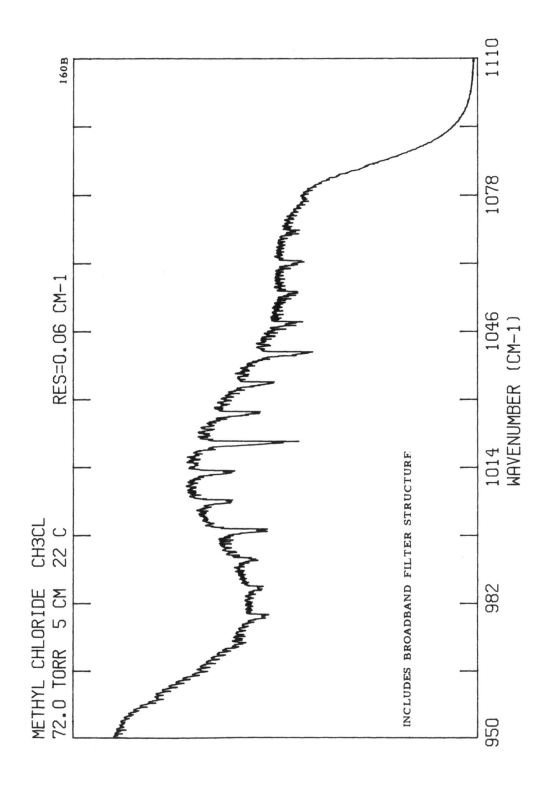

METHYL CHLORIDE   CH3CL
72.0 TORR   5 CM   22 C

RES=0.06 CM-1

160B

INCLUDES BROADBAND FILTER STRUCTURE

WAVENUMBER (CM-1)

950   982   1014   1046   1078   1110

METHYL CHLORIDE   CH3CL
72.0 TORR   5 CM   22 C

RES=0.06 CM-1

160B

WAVENUMBER (CM-1)

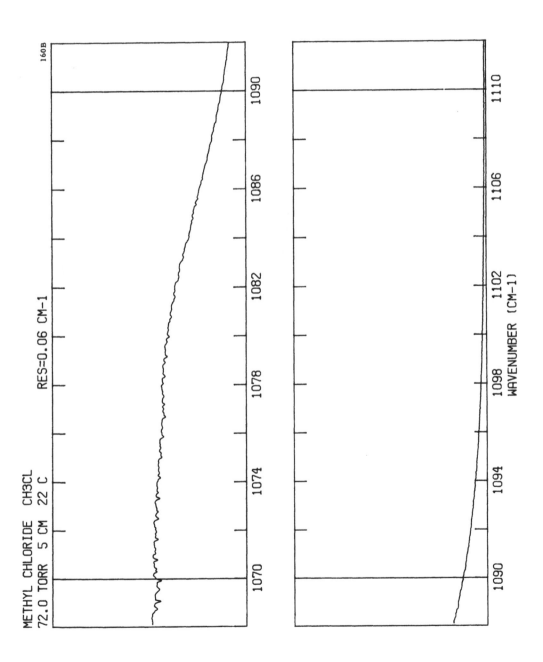

METHYL CHLORIDE   CH3CL
72.0 TORR   5 CM   22 C

RES=0.06 CM-1

160B

WAVENUMBER (CM-1)

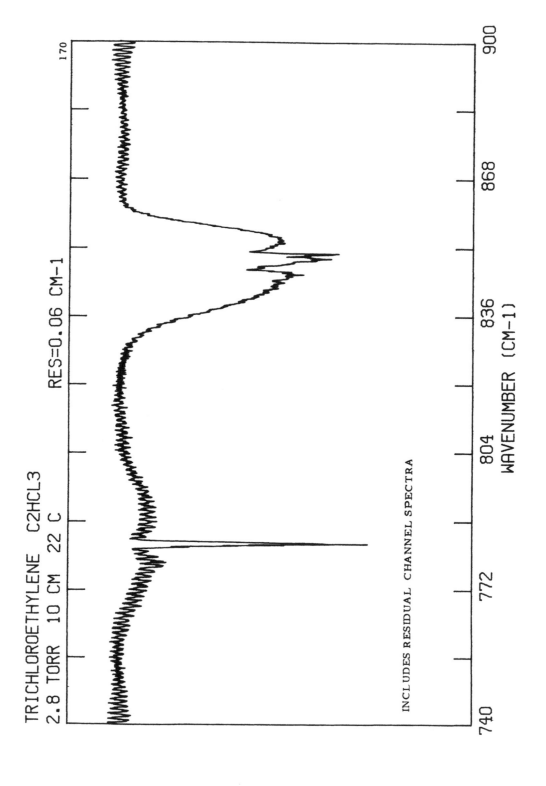

TRICHLOROETHYLENE C2HCL3

2.8 TORR 10 CM 22 C

RES=0.06 CM-1

170

INCLUDES RESIDUAL CHANNEL SPECTRA

WAVENUMBER (CM-1)

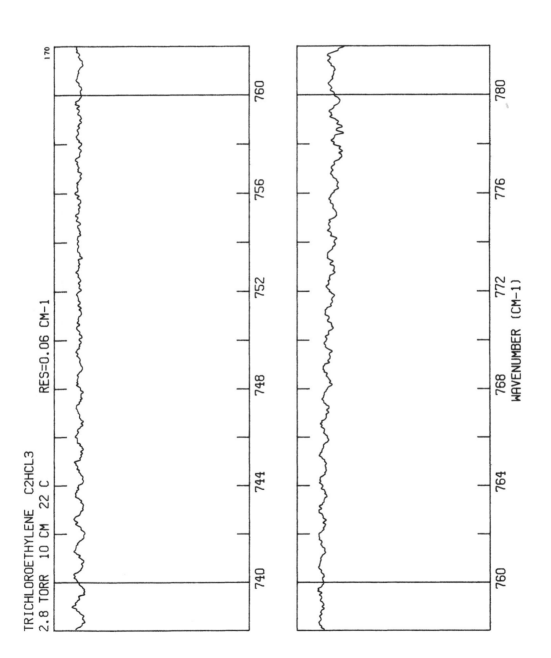

TRICHLOROETHYLENE C2HCL3

RES=0.06 CM-1

2.8 TORR 10 CM 22 C

WAVENUMBER (CM-1)

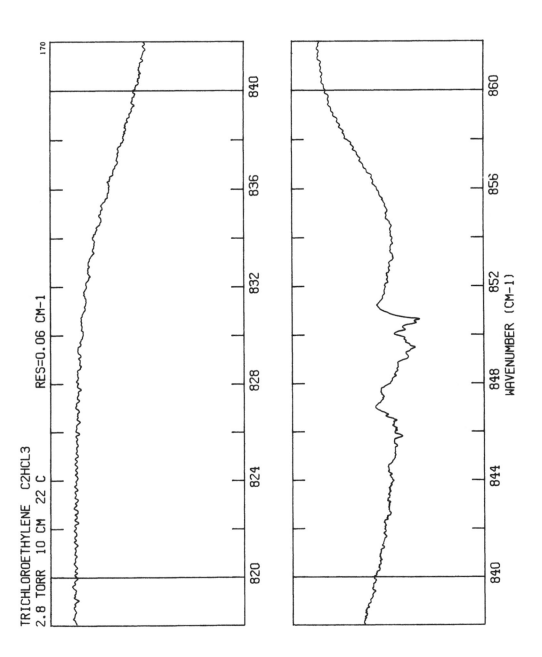

TRICHLOROETHYLENE C2HCL3
2.8 TORR 10 CM 22 C
RES=0.06 CM-1

WAVENUMBER (CM-1)

175

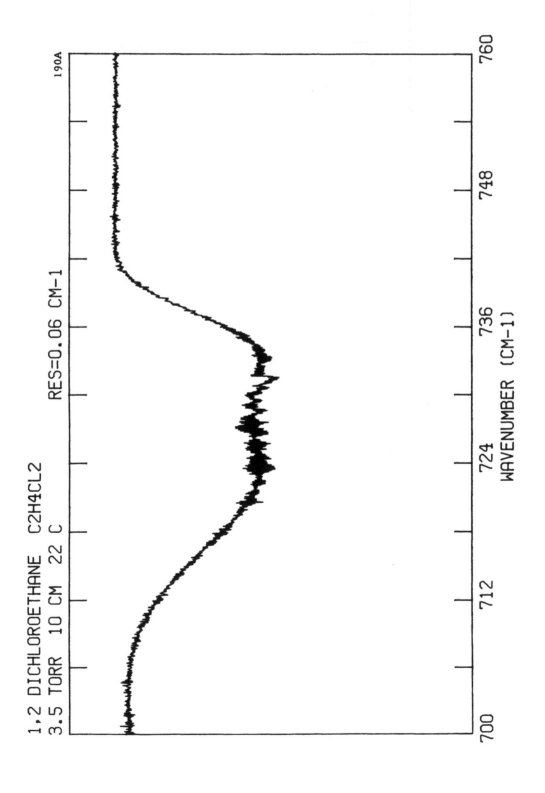

1,2 DICHLOROETHANE   C2H4CL2
3.5 TORR   10 CM   22 C

RES=0.06 CM-1

190A

WAVENUMBER (CM-1)

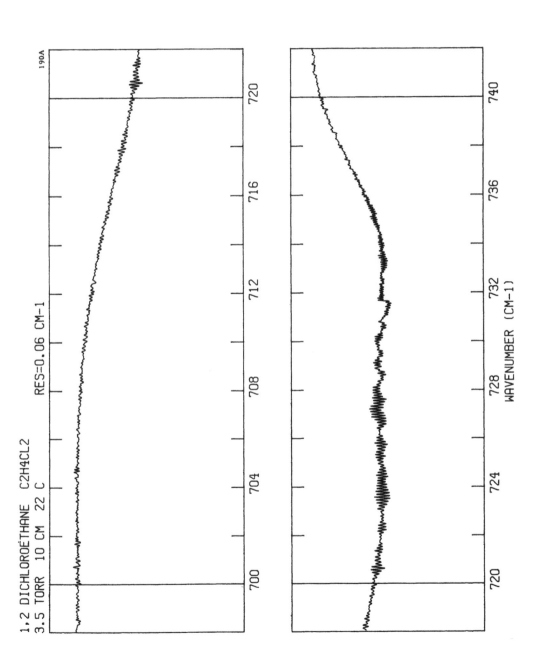

190A

1,2 DICHLOROETHANE C2H4CL2   RES=0.06 CM-1
3.5 TORR 10 CM 22 C

WAVENUMBER (CM-1)

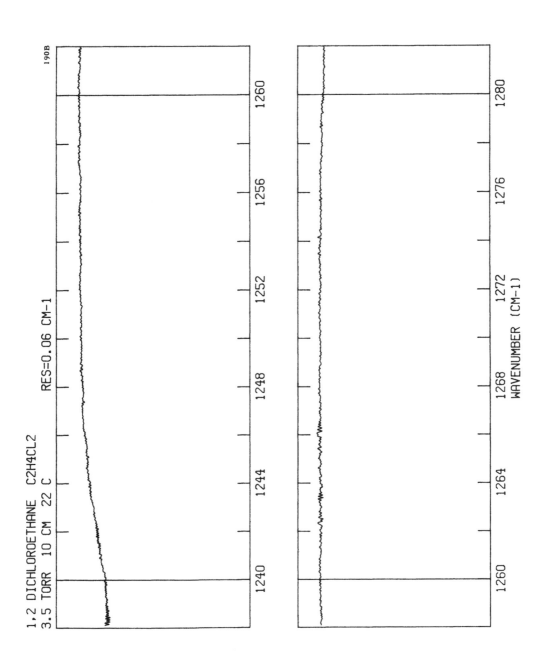

1,2 DICHLOROETHANE C2H4CL2

RES=0.06 CM-1

3.5 TORR 10 CM 22 C

190B

WAVENUMBER (CM-1)

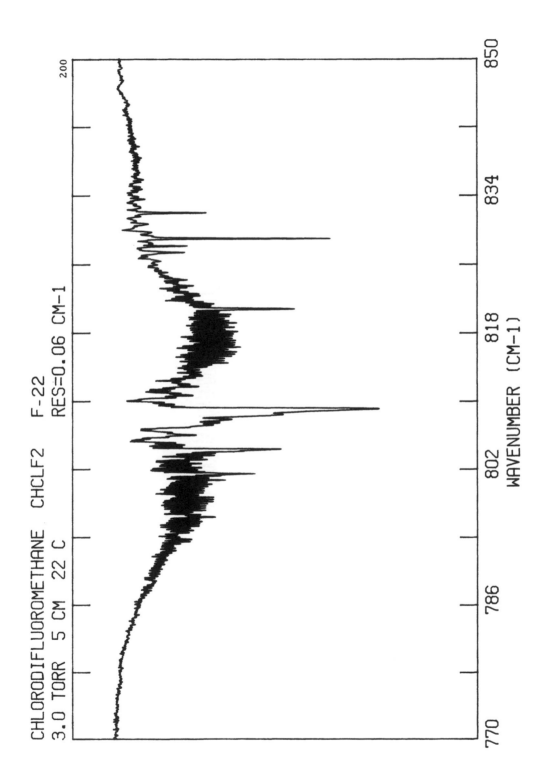

CHLORODIFLUOROMETHANE   CHCLF2   F-22
3.0 TORR   5 CM   22 C   RES=0.06 CM-1

WAVENUMBER (CM-1)

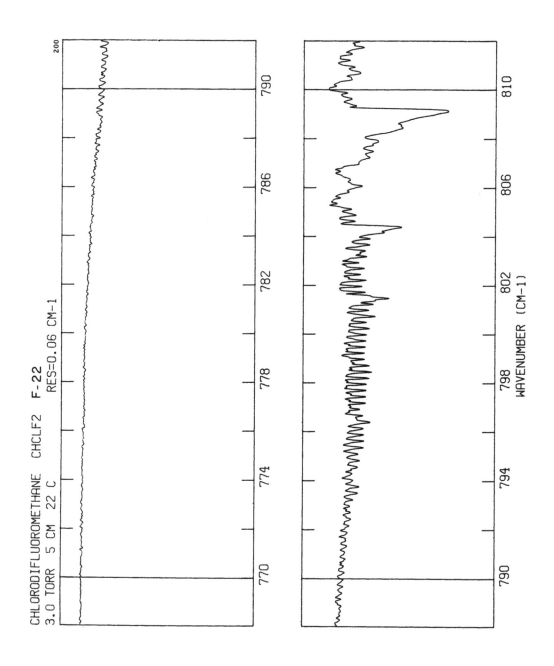

CHLORODIFLUOROMETHANE   CHCLF2   F-22
3.0 TORR   5 CM   22 C   RES=0.06 CM-1

WAVENUMBER (CM-1)

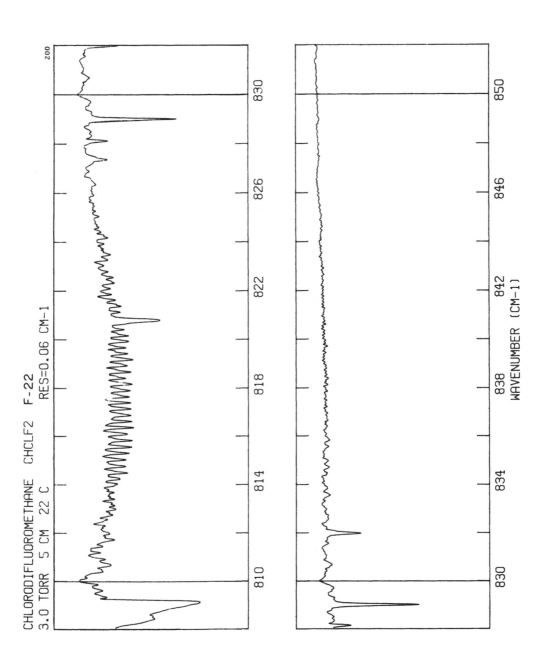

CHLORODIFLUOROMETHANE   CHCLF2   F-22   RES=0.06 CM-1
3.0 TORR 5 CM 22 C

WAVENUMBER (CM-1)

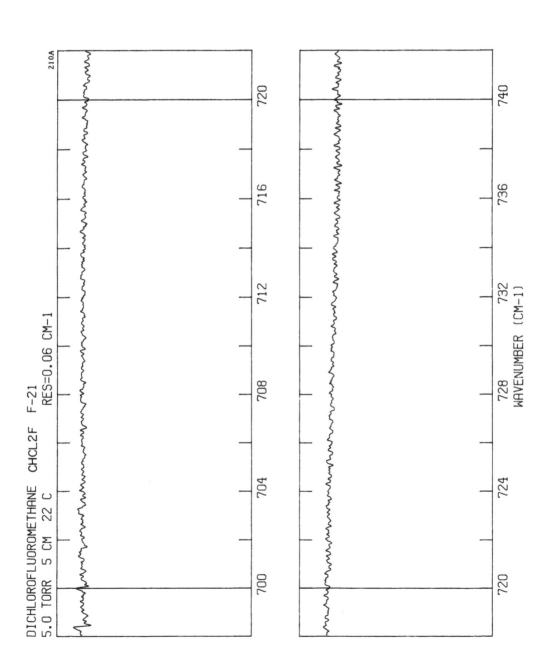

DICHLOROFLUOROMETHANE CHCL2F F-21
5.0 TORR 5 CM 22 C   RES=0.06 CM-1

210A

WAVENUMBER (CM-1)

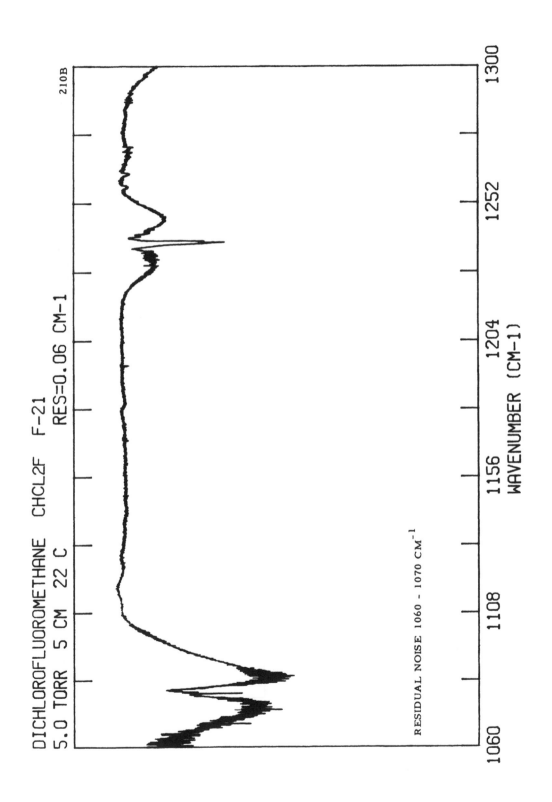

DICHLOROFLUOROMETHANE CHCL2F F-21 RES=0.06 CM-1

5.0 TORR 5 CM 22 C

210B

RESIDUAL NOISE 1060 - 1070 CM^{-1}

WAVENUMBER (CM-1)

1060   1108   1156   1204   1252   1300

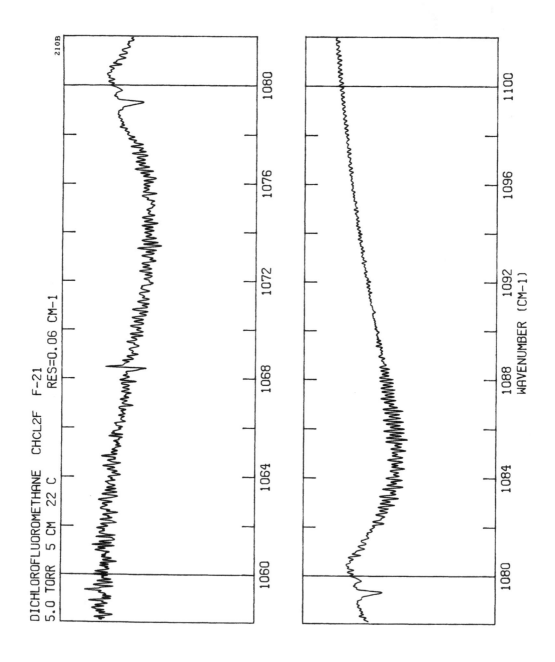

DICHLOROFLUOROMETHANE   CHCL2F   F-21   RES=0.06 CM-1
5.0 TORR   5 CM   22 C

WAVENUMBER (CM-1)

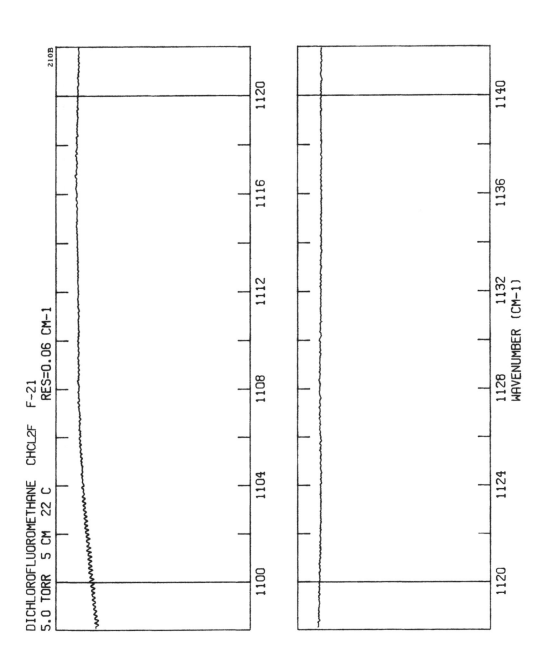

DICHLOROFLUOROMETHANE CHCL2F F-21
5.0 TORR 5 CM 22 C RES=0.06 CM-1

210B

1100   1104   1108   1112   1116   1120

1120   1124   1128   1132   1136   1140
WAVENUMBER (CM-1)

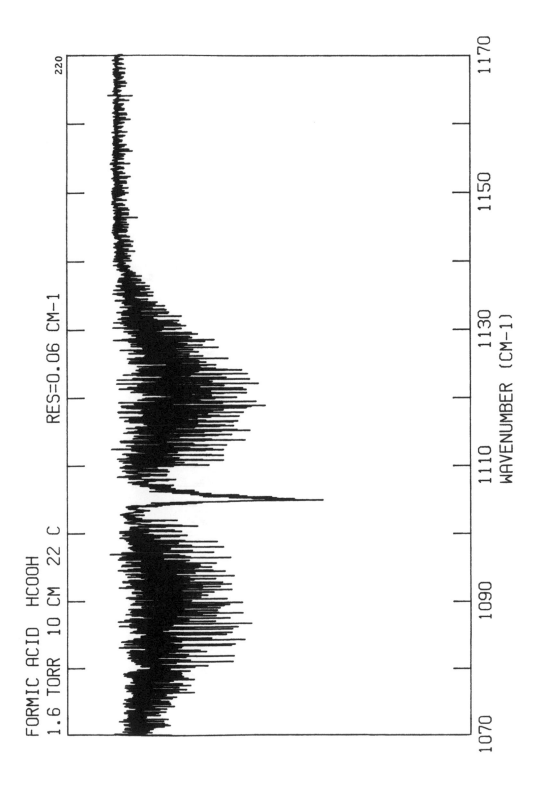

FORMIC ACID HCOOH
1.6 TORR 10 CM 22 C   RES=0.06 CM-1

220

WAVENUMBER (CM-1)

1070   1090   1110   1130   1150   1170

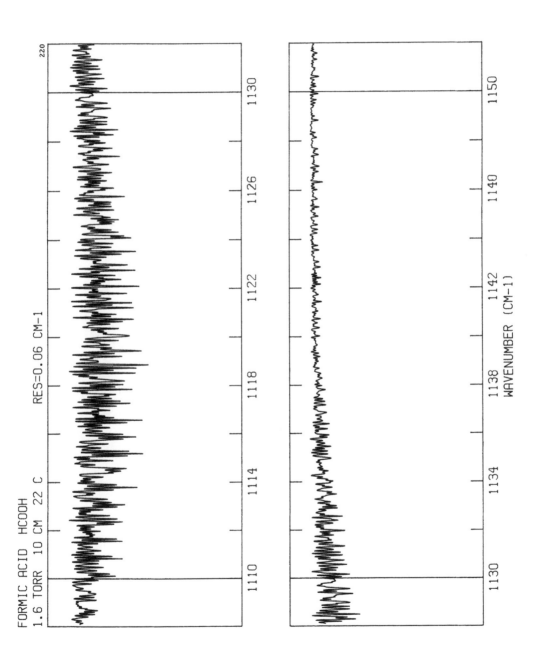

FORMIC ACID HCOOH
1.6 TORR 10 CM 22 C   RES=0.06 CM-1

WAVENUMBER (CM-1)

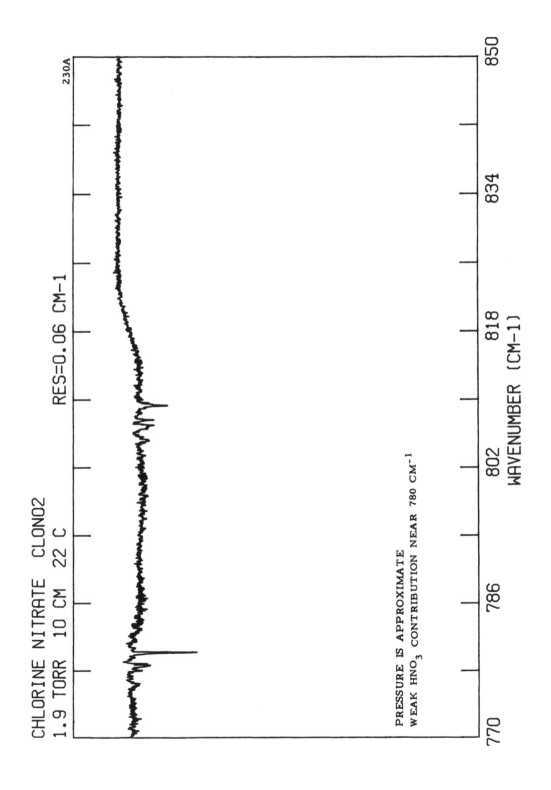

CHLORINE NITRATE   CLONO2
1.9 TORR   10 CM   22 C   RES=0.06 CM-1   230A

PRESSURE IS APPROXIMATE
WEAK HNO$_3$ CONTRIBUTION NEAR 780 CM^{-1}

WAVENUMBER (CM-1)

770   786   802   818   834   850

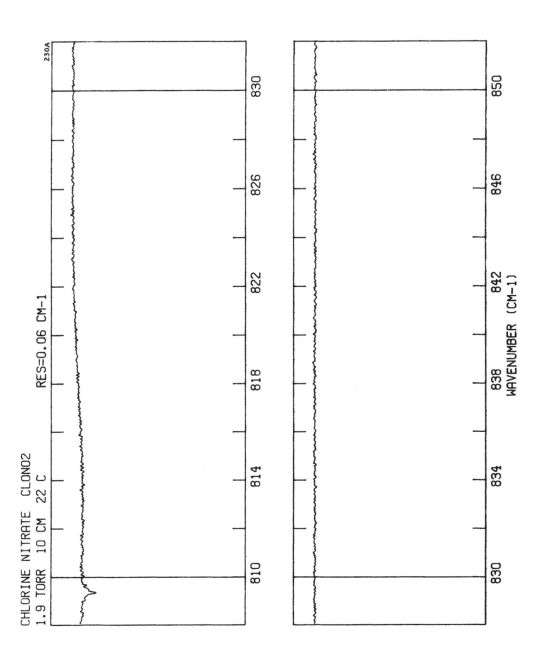

CHLORINE NITRATE CLONO2   RES=0.06 CM-1
1.9 TORR 10 CM 22 C

230A

WAVENUMBER (CM-1)

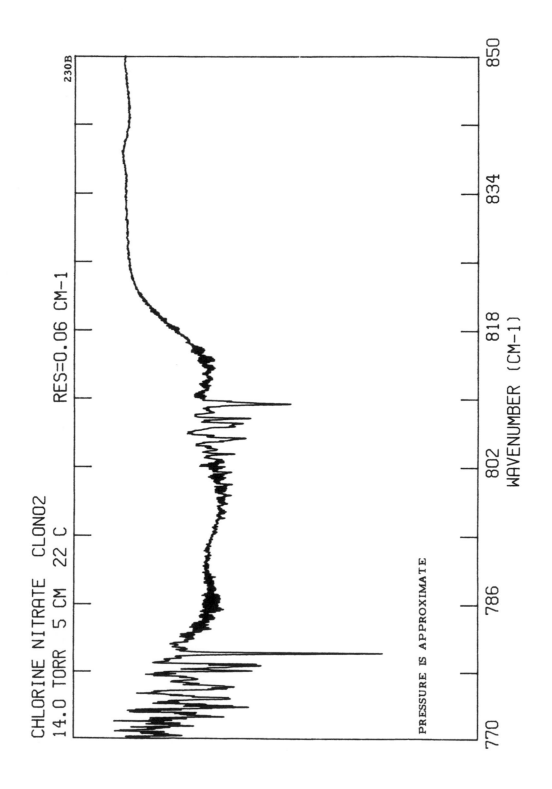

CHLORINE NITRATE CLONO2

14.0 TORR 5 CM 22 C

RES=0.06 CM-1

230B

WAVENUMBER (CM-1)

PRESSURE IS APPROXIMATE

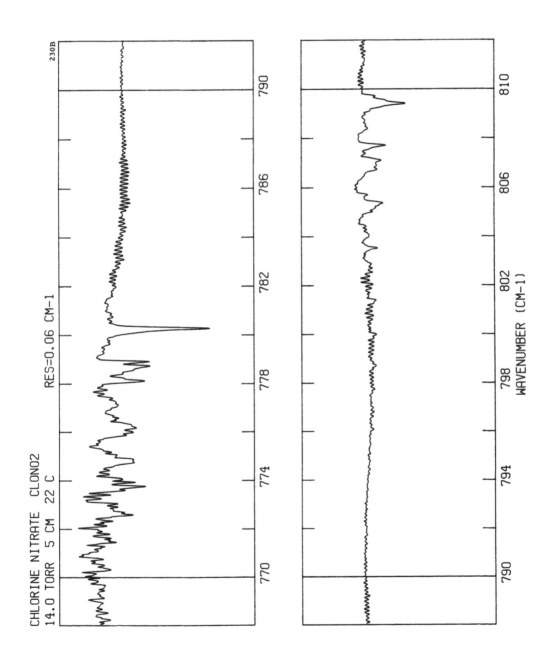

CHLORINE NITRATE CLONO2   RES=0.06 CM-1
14.0 TORR 5 CM 22 C

230B

WAVENUMBER (CM-1)

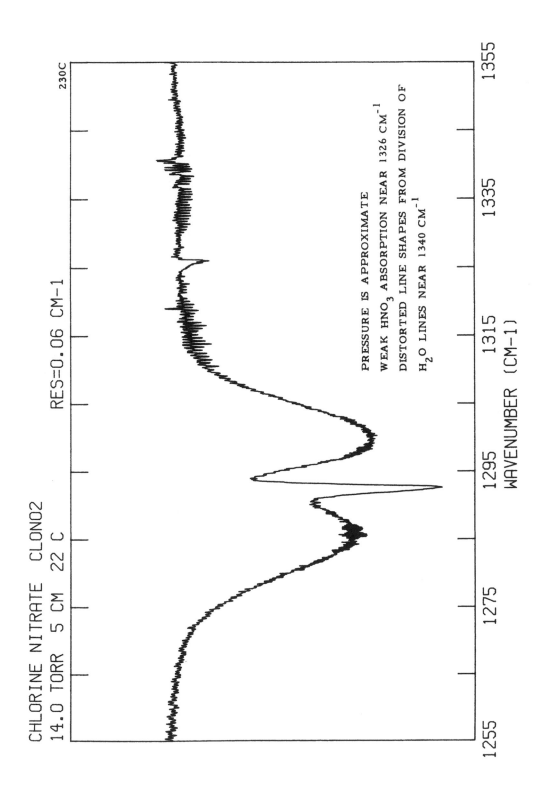

CHLORINE NITRATE CLONO2   RES=0.06 CM-1   230C
14.0 TORR 5 CM 22 C

PRESSURE IS APPROXIMATE
WEAK HNO$_3$ ABSORPTION NEAR 1326 CM^{-1}
DISTORTED LINE SHAPES FROM DIVISION OF
H$_2$O LINES NEAR 1340 CM^{-1}

WAVENUMBER (CM-1)

1255   1275   1295   1315   1335   1355

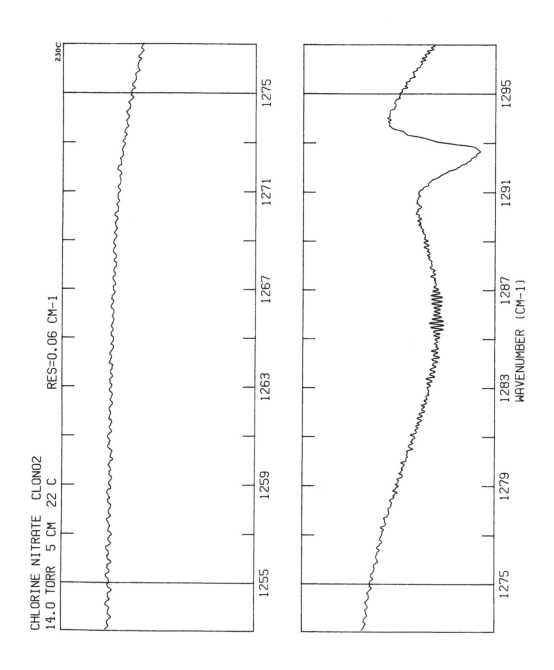

CHLORINE NITRATE   CLONO2
14.0 TORR   5 CM   22 C   RES=0.06 CM-1   230C

WAVENUMBER (CM-1)

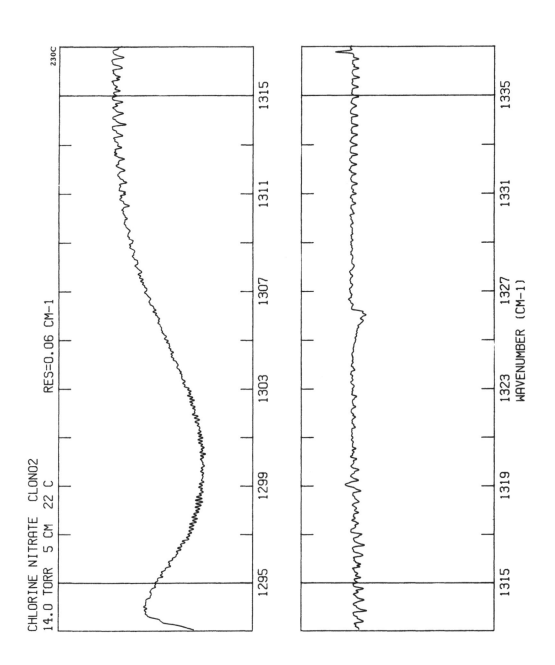

CHLORINE NITRATE CLONO2
14.0 TORR 5 CM 22 C

RES=0.06 CM-1

230C

1295 1299 1303 1307 1311 1315

1315 1319 1323 1327 1331 1335
WAVENUMBER (CM-1)

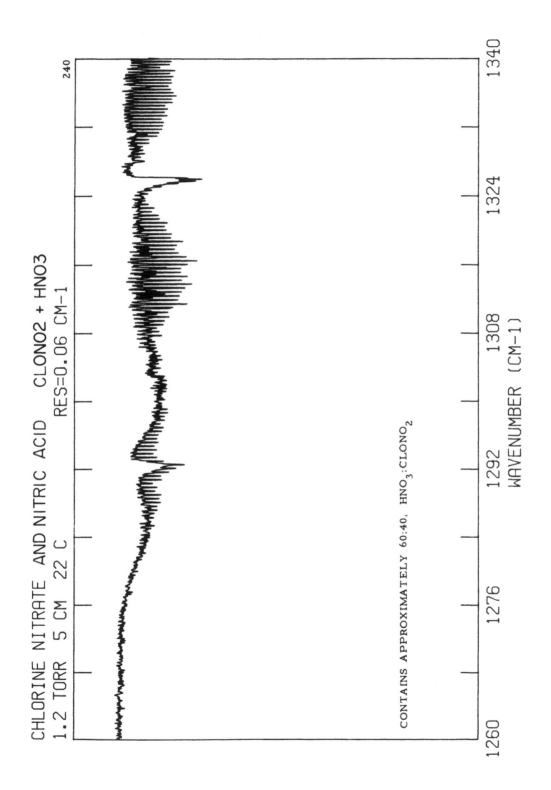

CHLORINE NITRATE AND NITRIC ACID   CLONO2 + HNO3
1.2 TORR   5 CM   22 C   RES=0.06 CM-1

240

WAVENUMBER (CM-1)

1260   1276   1292   1308   1324   1340

CONTAINS APPROXIMATELY 60:40, HNO_3:$CLONO_2$

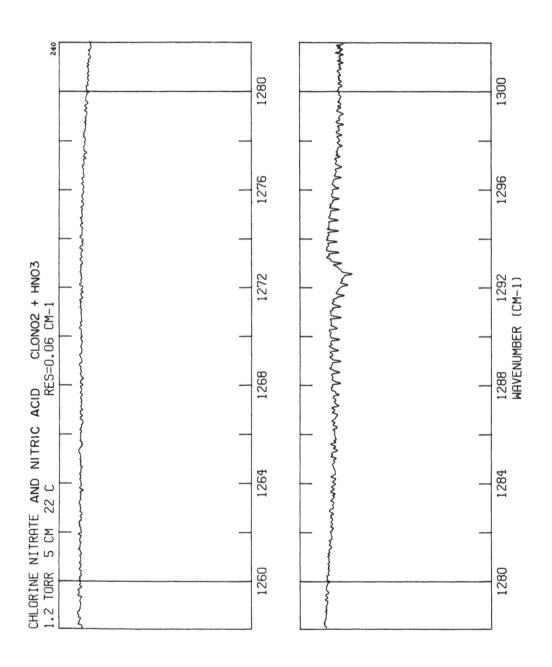

CHLORINE NITRATE AND NITRIC ACID   CLONO2 + HNO3
1.2 TORR   5 CM   22 C   RES=0.06 CM-1

WAVENUMBER (CM-1)

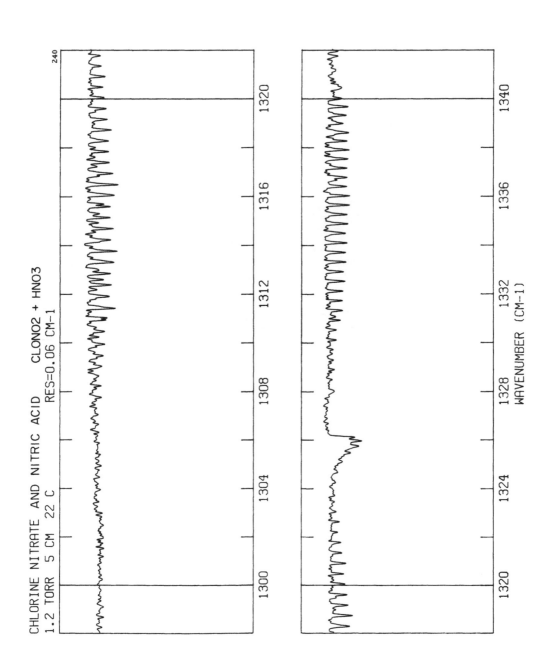

CHLORINE NITRATE AND NITRIC ACID   CLONO2 + HNO3
1.2 TORR   5 CM   22 C   RES=0.06 CM-1

WAVENUMBER (CM-1)

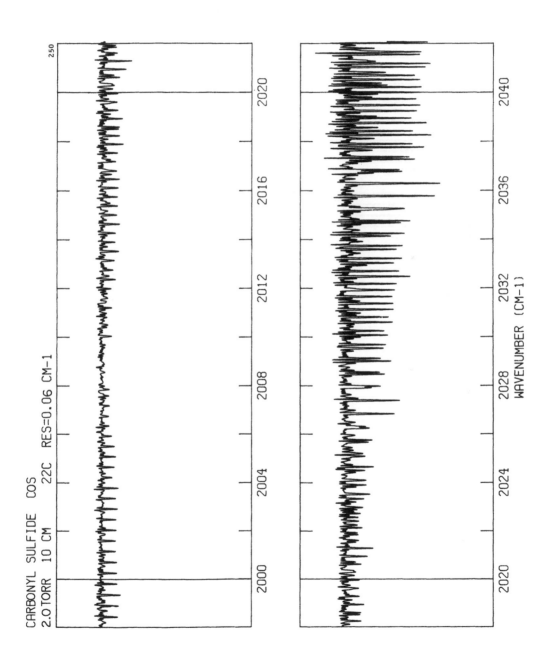

CARBONYL SULFIDE   COS   RES=0.06 CM-1
2.0 TORR   10 CM   22C

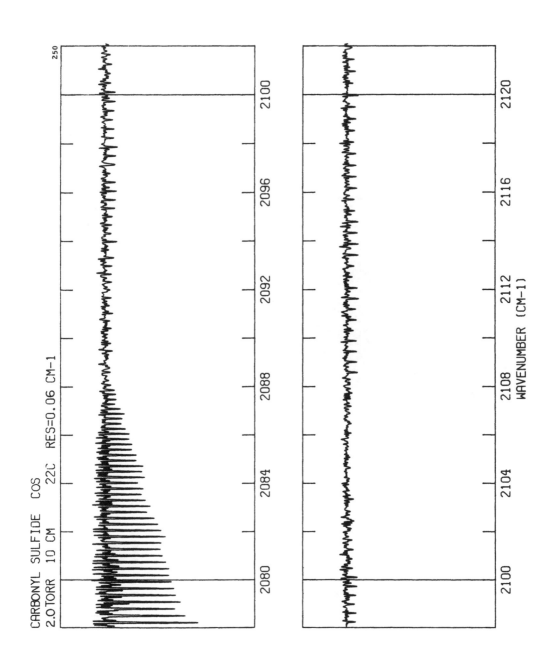

CARBONYL SULFIDE   COS
2.0 TORR   10 CM   22C   RES=0.06 CM-1

WAVENUMBER (CM-1)

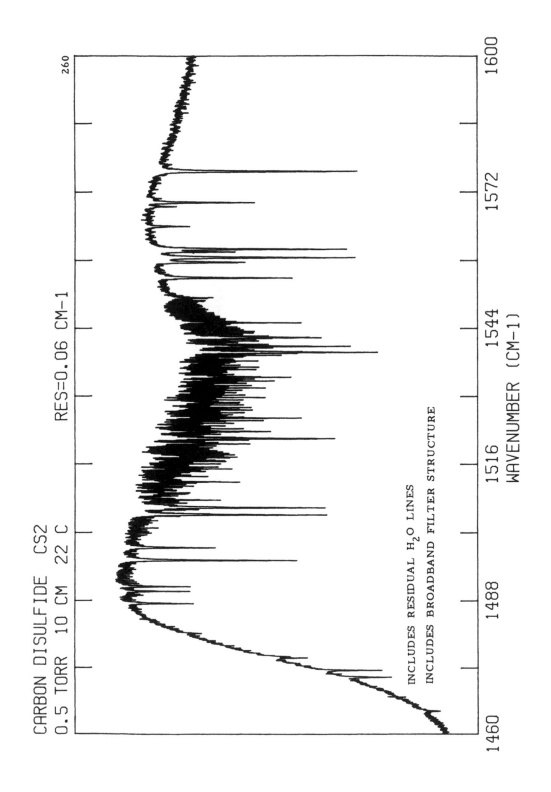

CARBON DISULFIDE CS2   RES=0.06 CM-1

0.5 TORR 10 CM 22 C

260

INCLUDES RESIDUAL H₂O LINES

INCLUDES BROADBAND FILTER STRUCTURE

WAVENUMBER (CM-1)

1460 1488 1516 1544 1572 1600

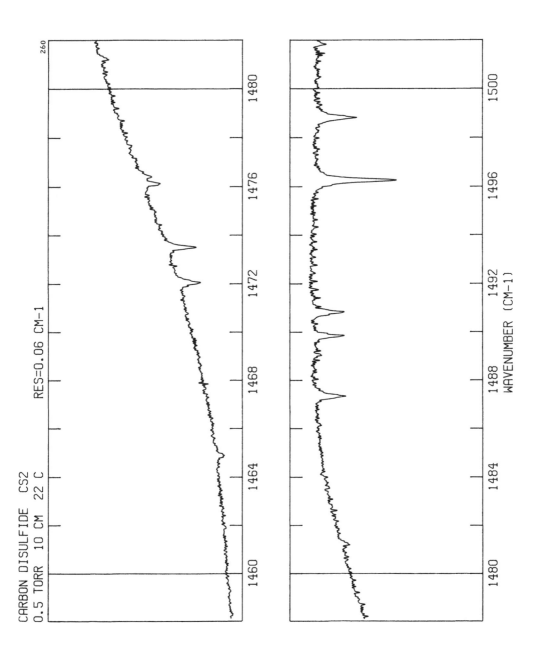

CARBON DISULFIDE   CS2

0.5 TORR   10 CM   22 C

RES=0.06 CM-1

260

WAVENUMBER (CM-1)

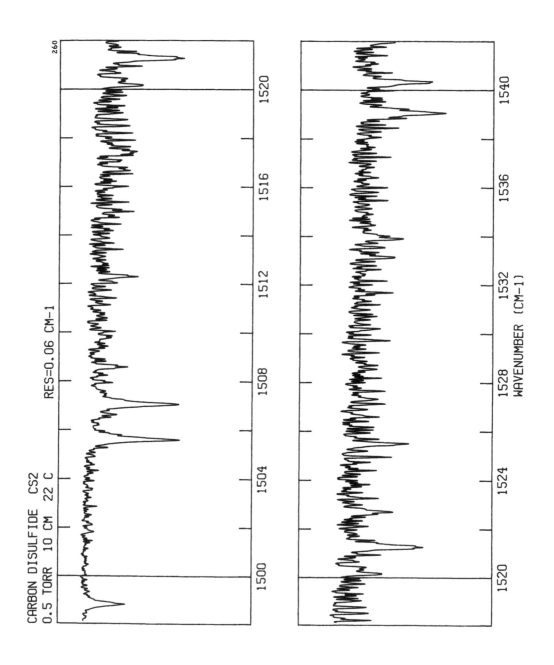

CARBON DISULFIDE CS2
0.5 TORR 10 CM 22 C   RES=0.06 CM-1

WAVENUMBER (CM-1)

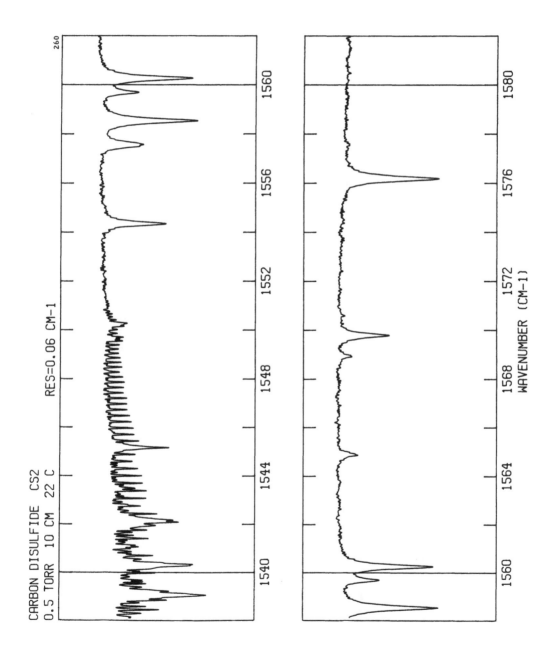

CARBON DISULFIDE CS2

0.5 TORR 10 CM 22 C

RES=0.06 CM-1

WAVENUMBER (CM-1)

CARBON DISULFIDE CS2
0.5 TORR 10 CM 22 C
RES=0.06 CM-1

WAVENUMBER (CM-1)

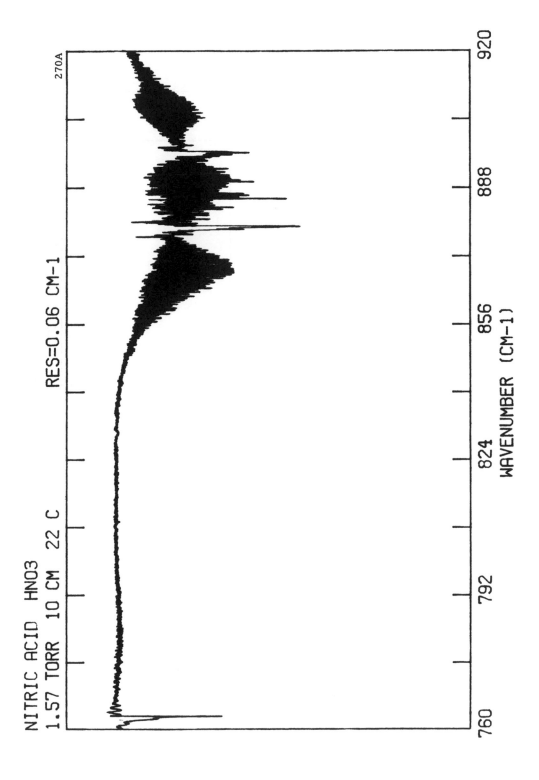

NITRIC ACID HNO3
1.57 TORR 10 CM 22 C

RES=0.06 CM-1

270A

WAVENUMBER (CM-1)

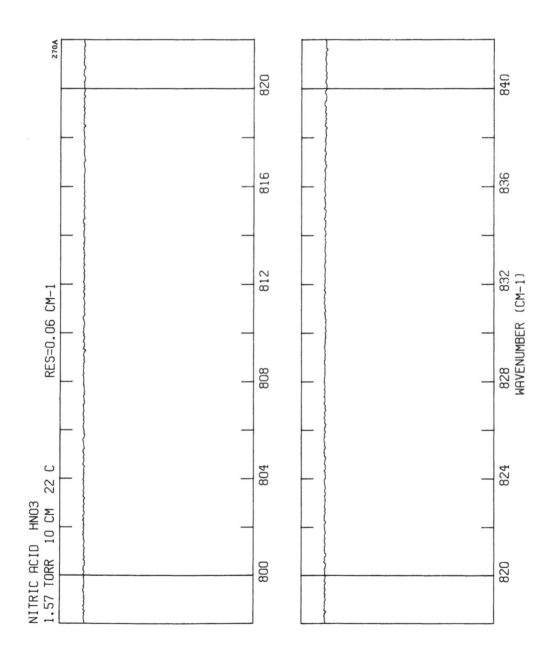

NITRIC ACID HNO3
1.57 TORR 10 CM 22 C

RES=0.06 CM-1

270A

WAVENUMBER (CM-1)

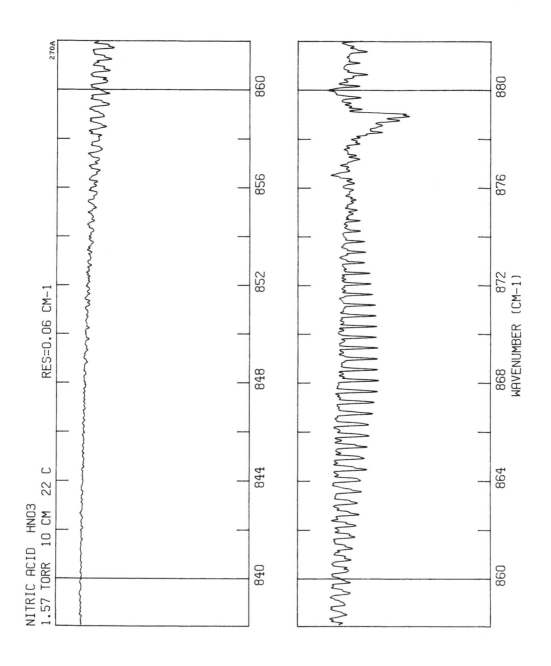

NITRIC ACID HNO3   RES=0.06 CM-1
1.57 TORR 10 CM 22 C

WAVENUMBER (CM-1)

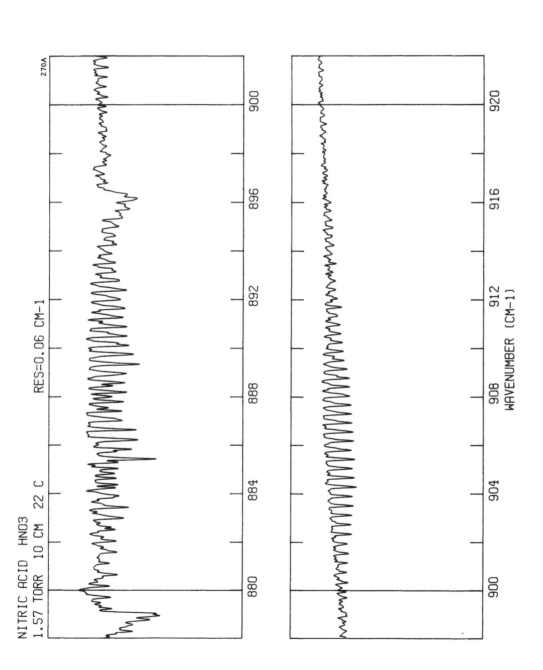

NITRIC ACID HNO3   RES=0.06 CM-1
1.57 TORR 10 CM 22 C

WAVENUMBER (CM-1)

229

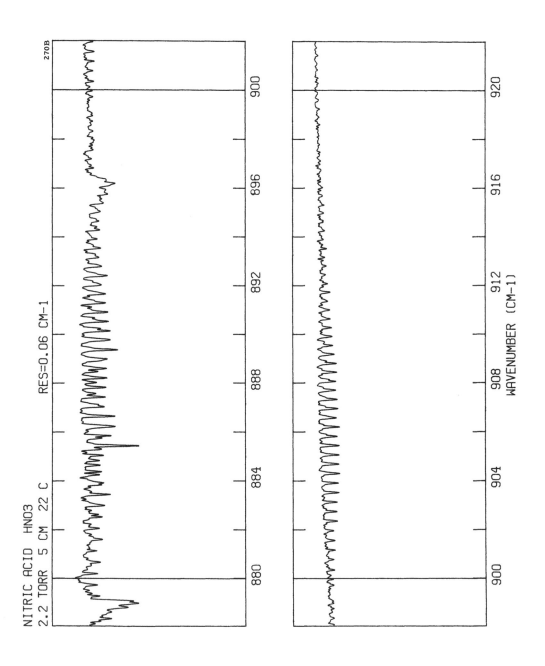

NITRIC ACID HNO3
2.2 TORR 5 CM 22 C

RES=0.06 CM-1

270B

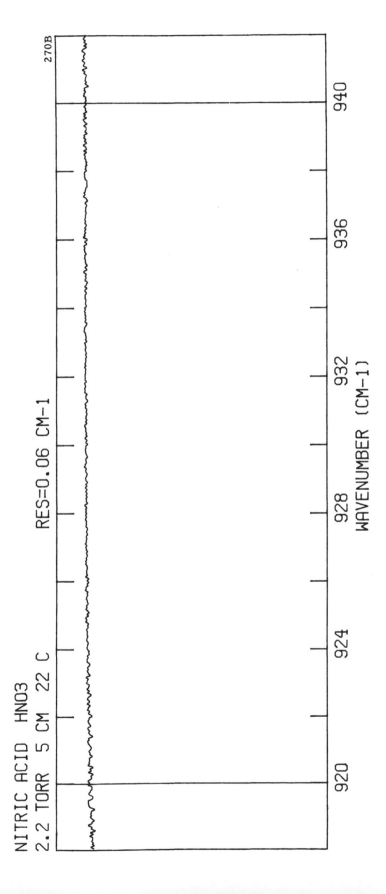

NITRIC ACID HNO3
2.2 TORR 5 CM 22 C

RES=0.06 CM-1

270B

WAVENUMBER (CM-1)

920 924 928 932 936 940

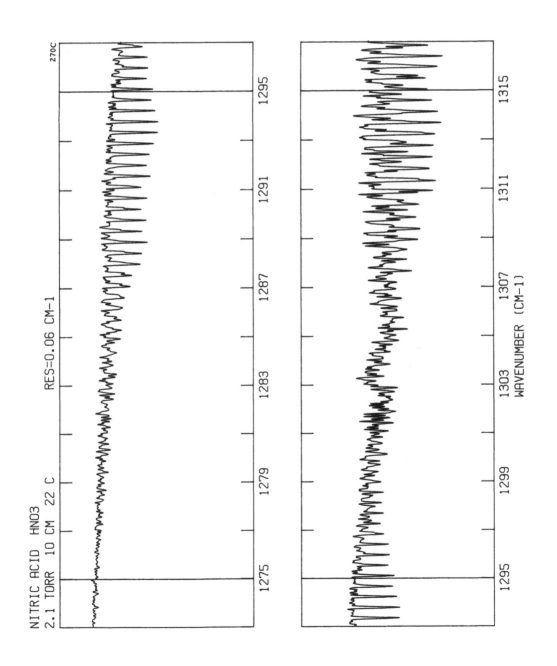

NITRIC ACID HNO3
2.1 TORR 10 CM 22 C   RES=0.06 CM-1

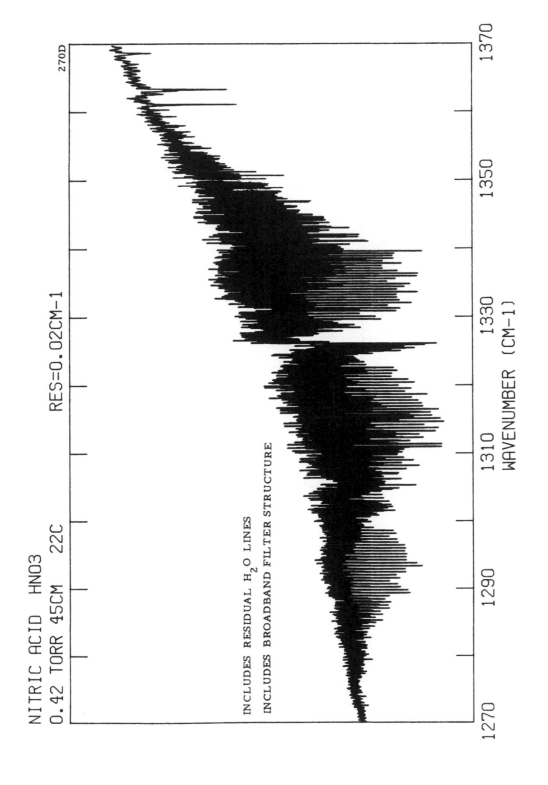

NITRIC ACID HNO3   RES=0.02CM-1

0.42 TORR 45CM 22C

INCLUDES RESIDUAL H₂O LINES

INCLUDES BROADBAND FILTER STRUCTURE

270D

WAVENUMBER (CM-1)

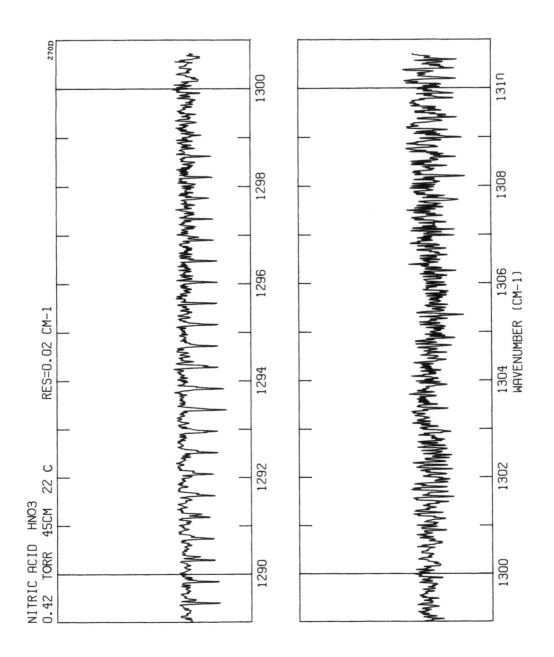

NITRIC ACID HNO3
0.42 TORR 45CM 22 C   RES=0.02 CM-1

270D

1300   1298   1296   1294   1292   1290

1310   1308   1306   1304   1302   1300
WAVENUMBER (CM-1)

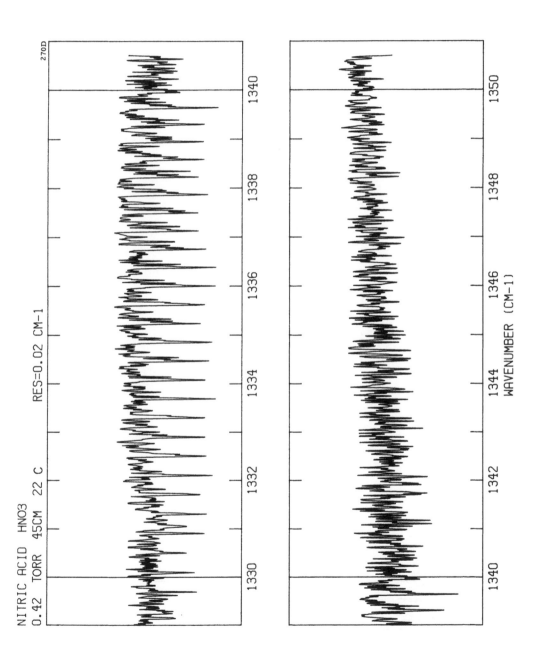

NITRIC ACID   HNO3   RES=0.02 CM-1
0.42 TORR   45CM   22 C

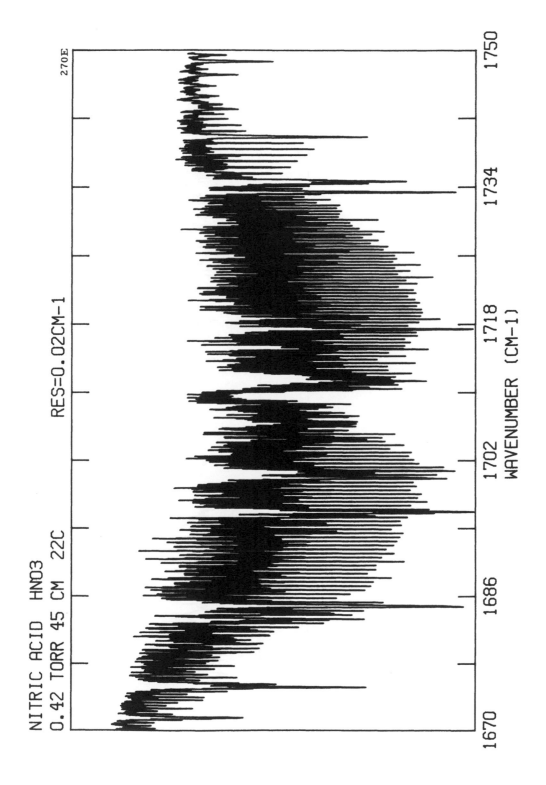

NITRIC ACID HNO3
0.42 TORR 45 CM 22C
RES=0.02CM-1
270E
WAVENUMBER (CM-1)

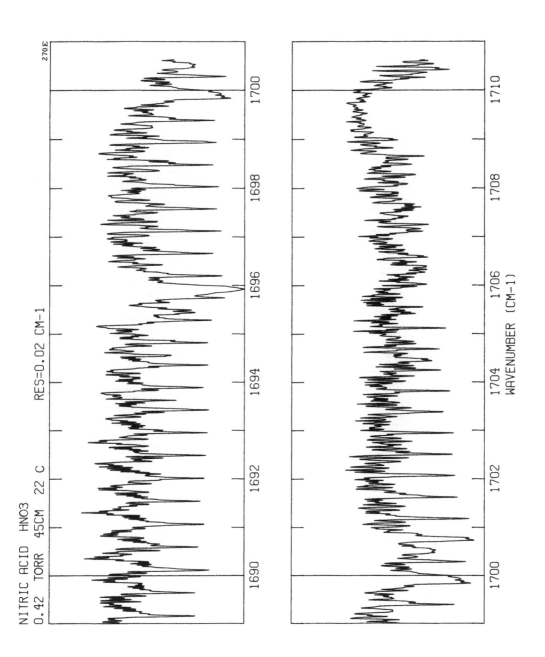

NITRIC ACID HNO3
0.42 TORR 45CM 22 C   RES=0.02 CM-1   270E

WAVENUMBER (CM-1)

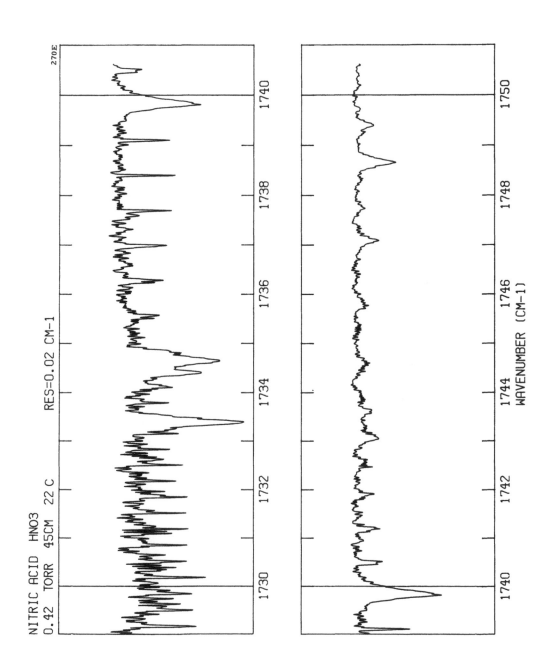

NITRIC ACID HNO3
0.42 TORR 45CM 22 C   RES=0.02 CM-1   270E

WAVENUMBER (CM-1)

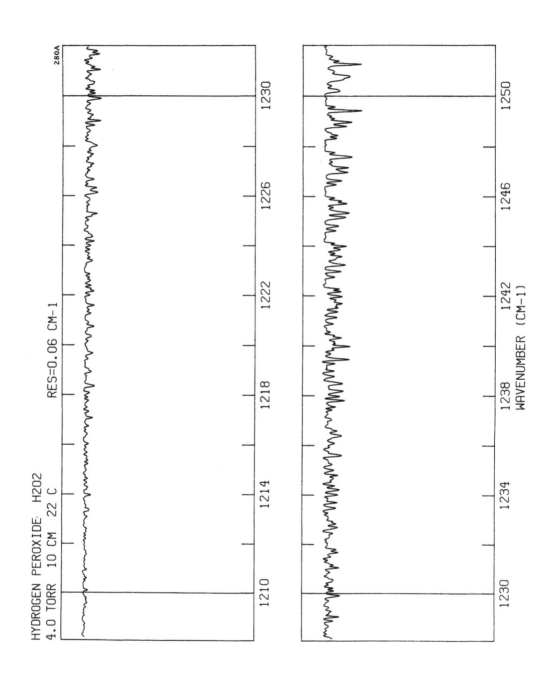

HYDROGEN PEROXIDE: H2O2

4.0 TORR   10 CM   22 C   RES=0.06 CM-1

WAVENUMBER (CM-1)

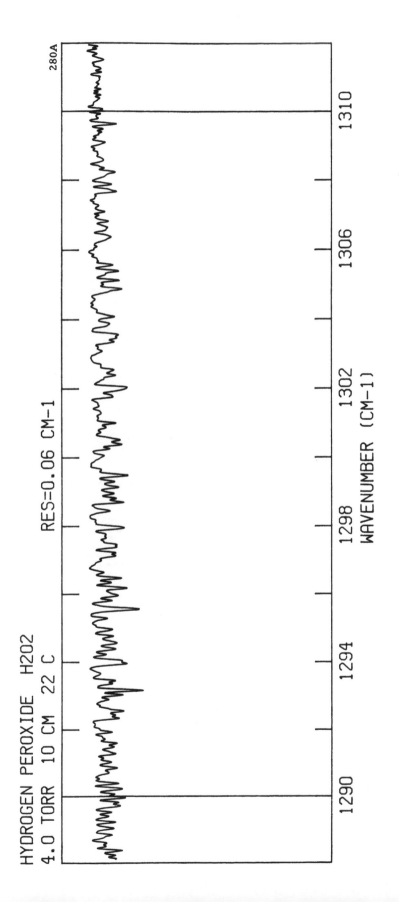

HYDROGEN PEROXIDE   H2O2
4.0 TORR   10 CM   22 C   RES=0.06 CM-1

280A

WAVENUMBER (CM-1)

1290   1294   1298   1302   1306   1310

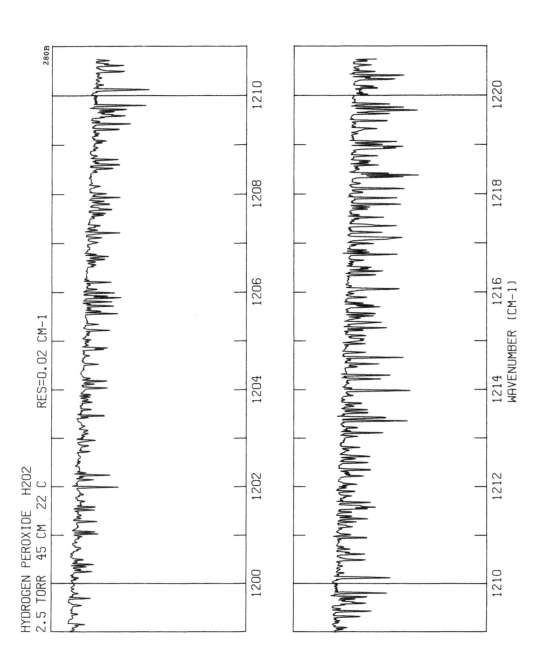

HYDROGEN PEROXIDE H2O2

2.5 TORR 45 CM 22 C   RES=0.02 CM-1

280B

WAVENUMBER (CM-1)

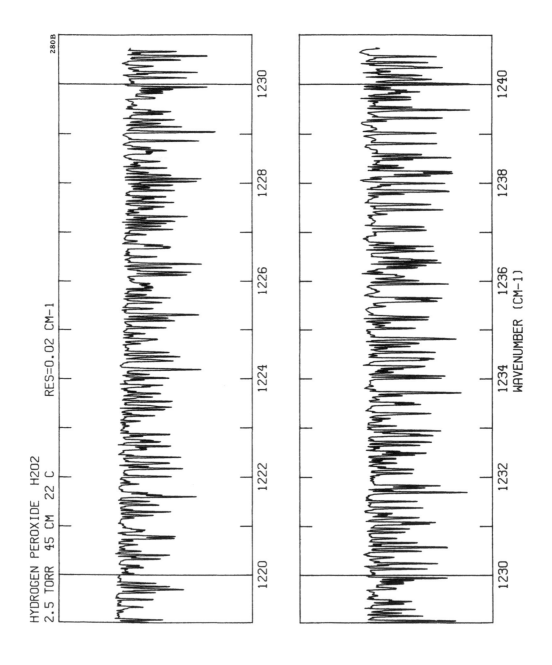

HYDROGEN PEROXIDE H2O2

2.5 TORR 45 CM 22 C   RES=0.02 CM-1

280B

WAVENUMBER (CM-1)

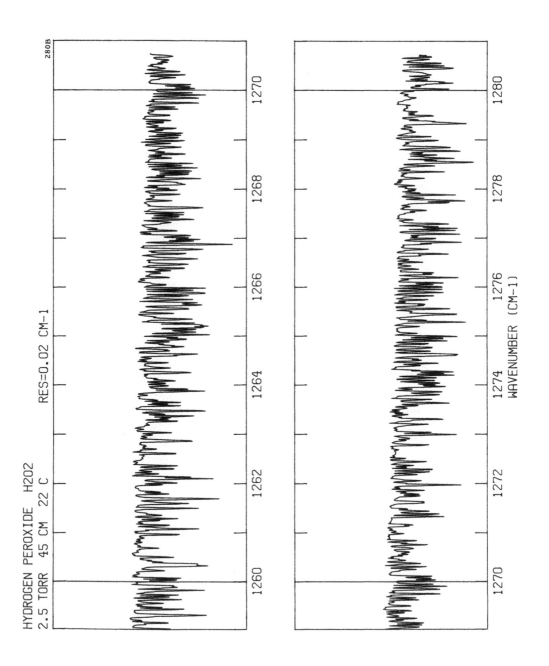

HYDROGEN PEROXIDE H2O2
2.5 TORR   45 CM   22 C   RES=0.02 CM-1

280B

WAVENUMBER (CM-1)

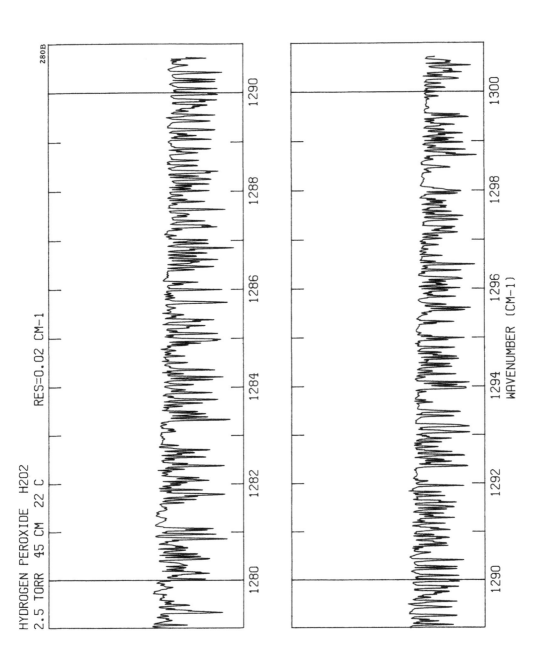

HYDROGEN PEROXIDE H2O2

2.5 TORR 45 CM 22 C

RES=0.02 CM-1

280B

1290 1288 1286 1284 1282 1280

1300 1298 1296 1294 1292 1290

WAVENUMBER (CM-1)

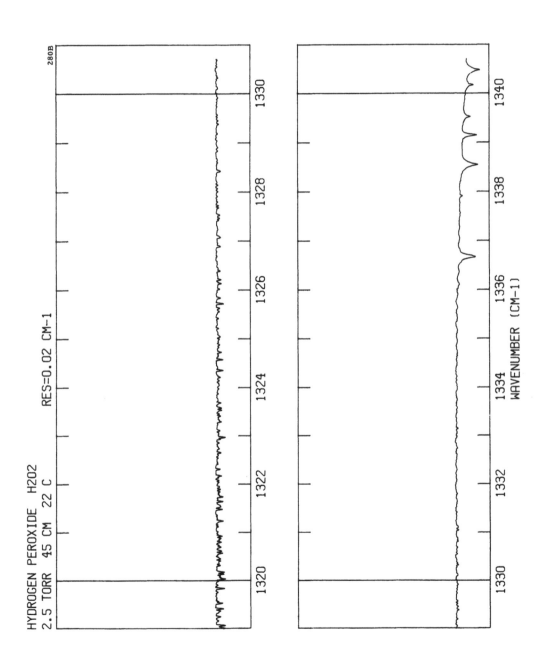

HYDROGEN PEROXIDE H2O2

2.5 TORR 45 CM 22 C

RES=0.02 CM-1

280B

WAVENUMBER (CM-1)

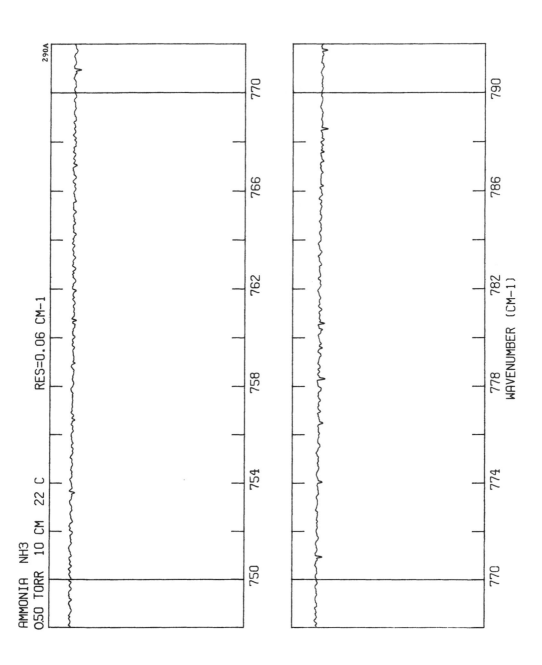

AMMONIA NH3

0.50 TORR 10 CM 22 C

RES=0.06 CM-1

290A

WAVENUMBER (CM-1)

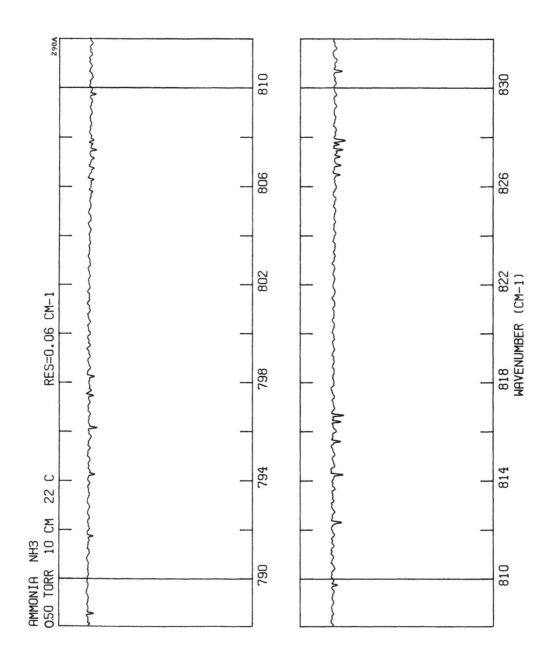

AMMONIA NH3
0.50 TORR 10 CM 22 C   RES=0.06 CM-1

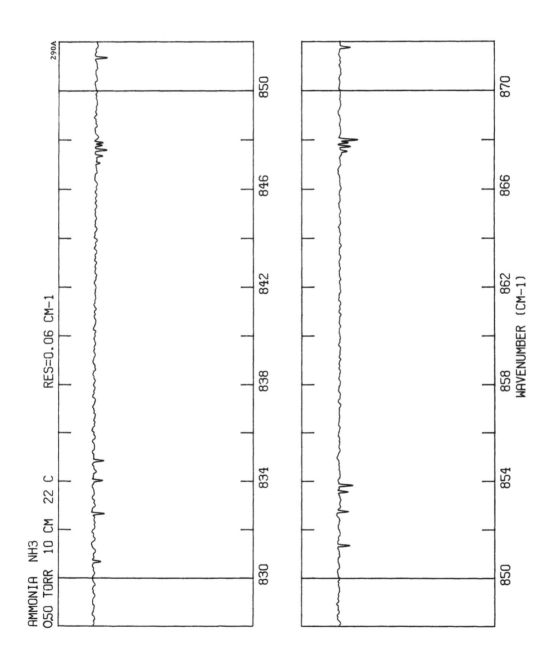

AMMONIA   NH3
O50 TORR   10 CM   22 C   RES=0.06 CM-1   290A

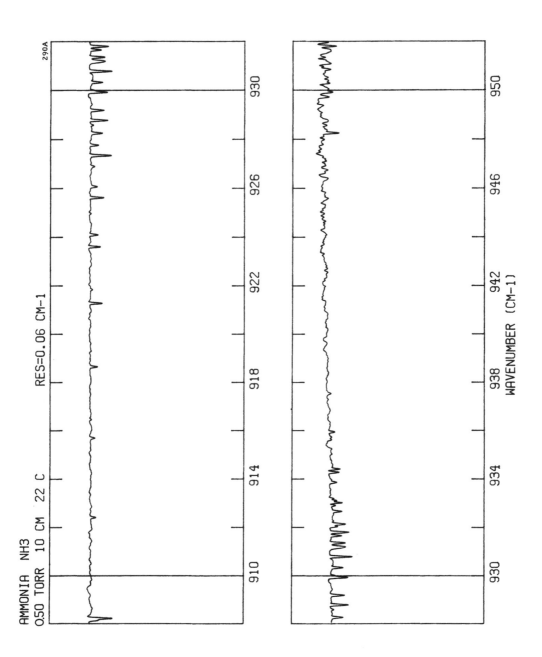

AMMONIA NH3
0.50 TORR 10 CM 22 C

RES=0.06 CM-1

290A

WAVENUMBER (CM-1)

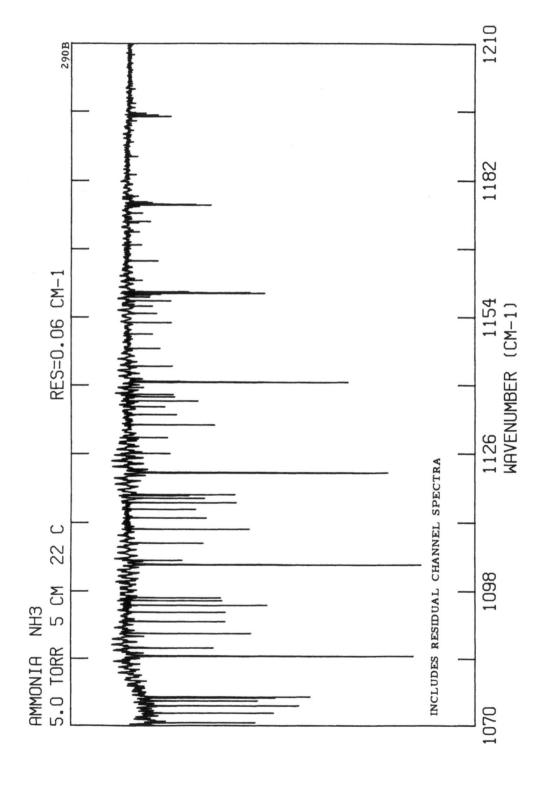

AMMONIA   NH3
5.0 TORR   5 CM   22 C   RES=0.06 CM-1   290B

INCLUDES RESIDUAL CHANNEL SPECTRA

WAVENUMBER (CM-1)

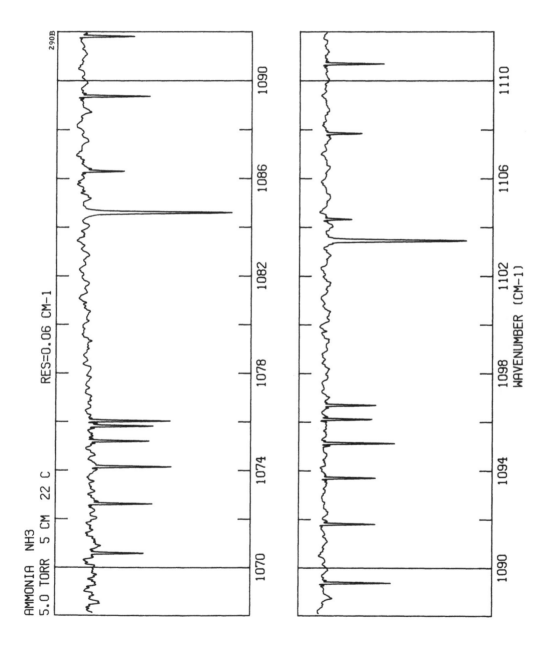

AMMONIA NH3   RES=0.06 CM-1
5.0 TORR 5 CM 22 C

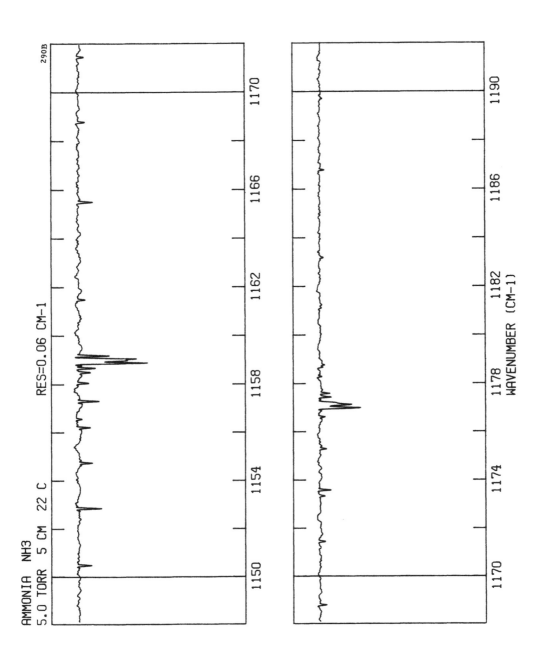

AMMONIA NH3
5.0 TORR 5 CM 22 C
RES=0.06 CM-1
290B

1150 1154 1158 1162 1166 1170

1170 1174 1178 1182 1186 1190
WAVENUMBER (CM-1)

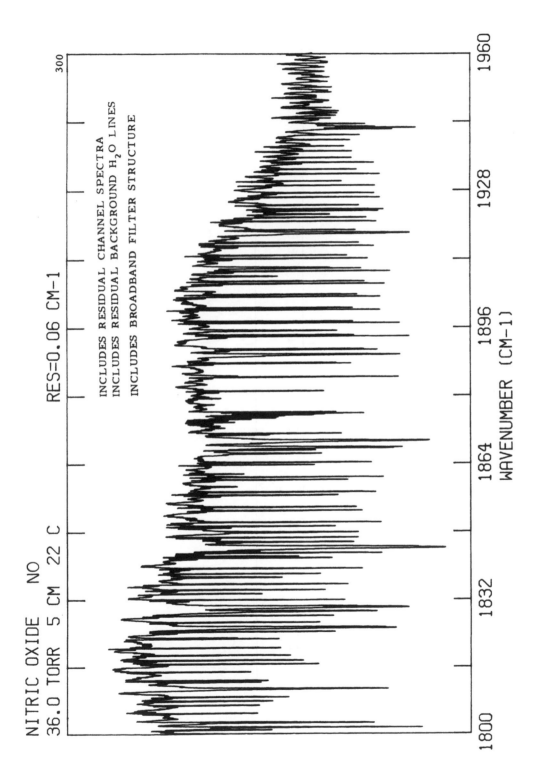

NITRIC OXIDE   NO
36.0 TORR   5 CM   22 C

RES=0.06 CM-1

INCLUDES RESIDUAL CHANNEL SPECTRA
INCLUDES RESIDUAL BACKGROUND H_2O LINES
INCLUDES BROADBAND FILTER STRUCTURE

WAVENUMBER (CM-1)

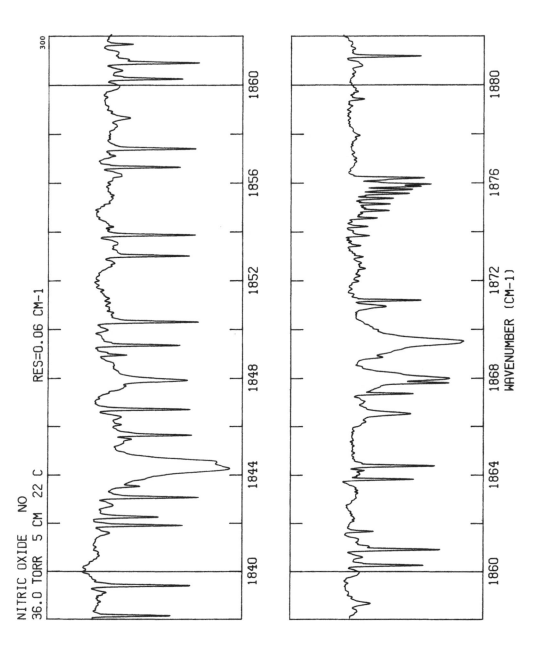

NITRIC OXIDE   NO
36.0 TORR 5 CM 22 C

RES=0.06 CM-1

WAVENUMBER (CM-1)

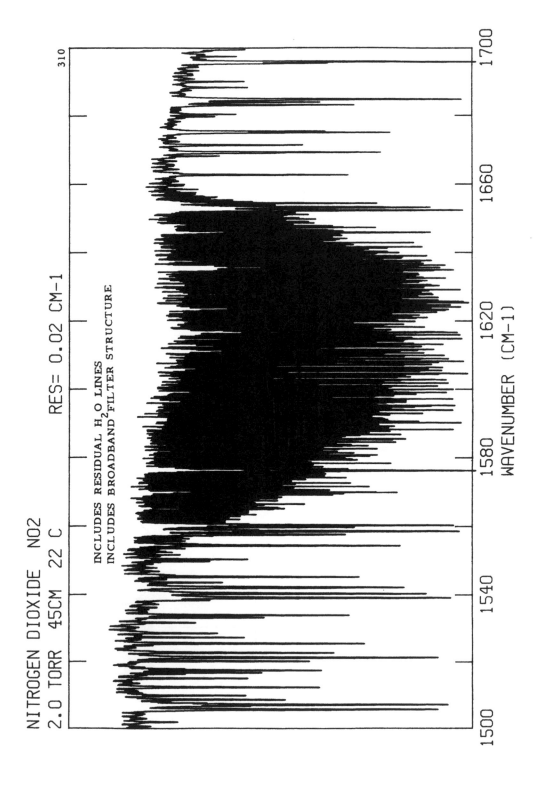

NITROGEN DIOXIDE   NO2

2.0 TORR   45CM   22 C

RES= 0.02 CM-1

INCLUDES RESIDUAL H₂O LINES

INCLUDES BROADBAND FILTER STRUCTURE

310

WAVENUMBER (CM-1)

1500   1540   1580   1620   1660   1700

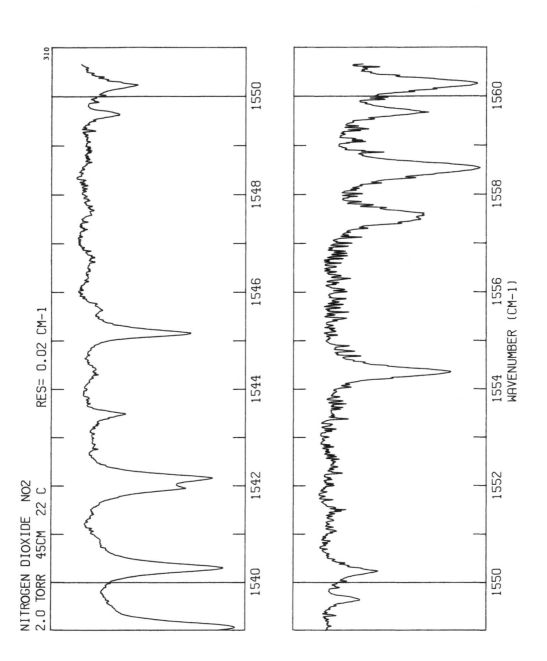

NITROGEN DIOXIDE   NO2

2.0 TORR   45CM   22 C

RES= 0.02 CM-1

310

WAVENUMBER (CM-1)

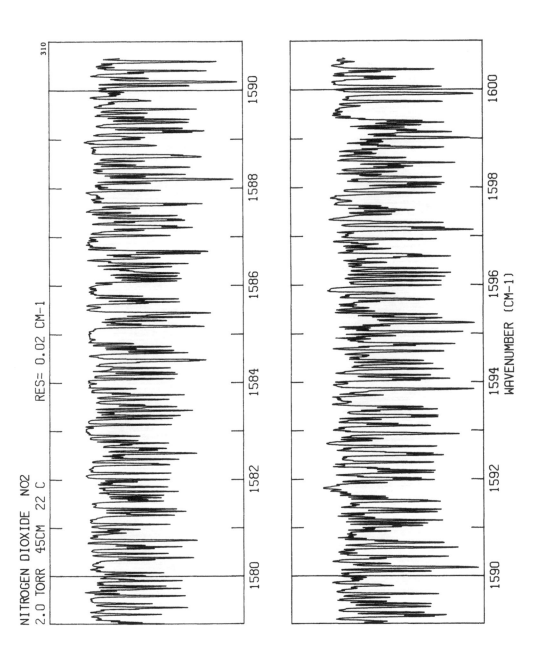

NITROGEN DIOXIDE NO2

2.0 TORR 45CM 22 C

RES= 0.02 CM-1

WAVENUMBER (CM-1)

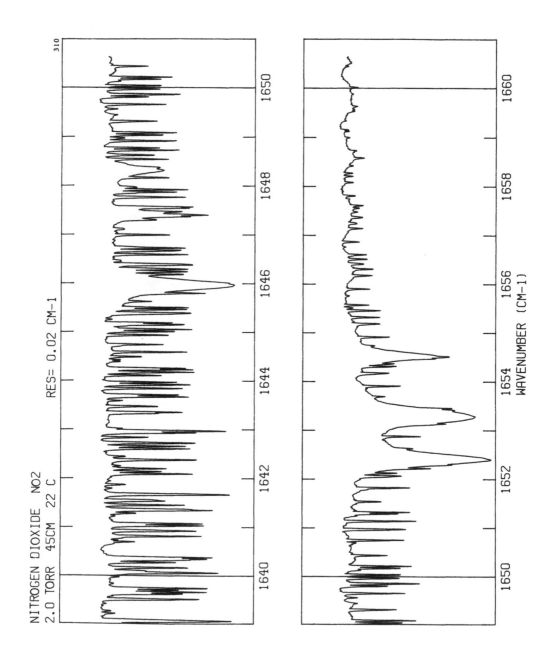

NITROGEN DIOXIDE NO2
2.0 TORR 45CM 22 C   RES= 0.02 CM-1

WAVENUMBER (CM-1)

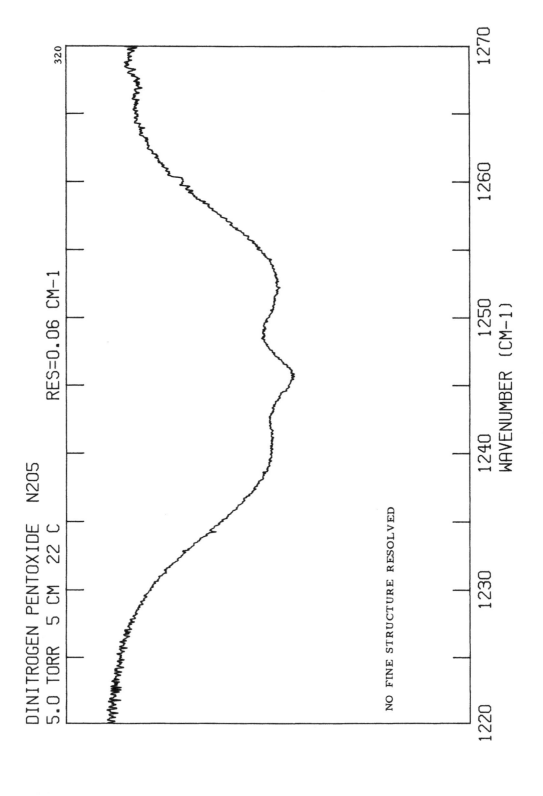

DINITROGEN PENTOXIDE   N2O5
5.0 TORR   5 CM   22 C   RES=0.06 CM-1

320

NO FINE STRUCTURE RESOLVED

WAVENUMBER (CM-1)

Milton Keynes UK
Ingram Content Group UK Ltd.
UKHW051933141024
449569UK00027B/1468